Scope of Multipolar MMC HVDC transmission Dynamics

By
Rowan Theni

Contents

Chapter 1

Introduction

1.1 General

Electricity is the backbone of modern human civilisation. The electrical power transmission networks connect the generators to the end users to fulfil the ever-growing electrical energy requirement. Conventionally the high-voltage ac transmission was the choice for the power transmission network. However, it suffers from many drawbacks. In recent decades, high voltage dc (HVDC) has emerged as a new and effective alternative for bulk power transmission over a long distance [1,2]. Because of global warming and other environmental concerns, the whole power industry is moving toward solar, wind, and other renewable energy sources. Furthermore, electricity is now being used as a fuel for transportation services. All of these developments make HVDC transmission more relevant [3]. The HVDC transmission requires converters at both terminals. These converters can be a current source converter (CSC) or a voltage source converter (VSC). The most modern and cutting-edge converter for VSC-based HVDC is the modular multilevel converter (MMC) [4–9]. Recent developments in HVDC breaker technology give MMC-based HVDC systems as an additional boost [10]. The submodule-based MMC topology outperforms alternative converter topologies due to a variety of technological and economic advantages.

The history of MMC can be traced back to the mid-1970s. Three- and five-level inverters, as well as the first patent capable of generating multiple voltages from various dc voltage sources, are developed by connecting single-phase inverters in series [11, 12]. Initial pioneering work on "modular multilevel structure" includes a single-phase circuit for plasma stabilization and a three-phase circuit for motor drive [13, 14]. However, the first time MMC was proposed in German patent by Marquardt in the year 2002. [7]. The circuit configuration and operating principle were described in [15–17].

The MMC technology has evolved remarkably in the recent decade [18–21]. MMC is the most optimal linear time-invariant converter system in terms of arm-current distribution [22]. Various advantages of the MMC comprise modularity, scalability, reduced losses, high power quality, reduced operating cost, reduced cost of maintenance, and filter-less and transformerless operations. These advantages make the MMC a suitable candidate for high voltage and high power applications [3,23,24]. The sophisticated control that considers concurrent regulation of (i) power flow, (ii) dc voltage control, (iii) eliminating the circulating current, (iv) capacitor voltage balancing, (v) active-reactive power decoupling, (vi) dc-fault handling, etc., is a major challenge in MMC operation [25]. Furthermore, the efficiency is heavily dependent on modulation, and submodule selection techniques [26]. Several control schemes have been developed to operate the MMC for a variety of applications [27–35].

The main focus of this thesis is enabling the MMC for efficient operation of the HVDC transmission. Various submodule topologies have been investigated. The significance of the circulating current controller has been examined. New controllers have been proposed to assist power-decoupling, transformerless operation, and multipolar operation.

1.2 Growth of VSC HVDC Transmission

Traditionally the power transmission was dependent on the high-voltage ac only. However, with new innovations, global power transmission is rapidly moving towards high-voltage dc transmission. In the upcoming years, the compound average growth rate of the HVDC is predicted to be 11% [36]. This momentum is due to the

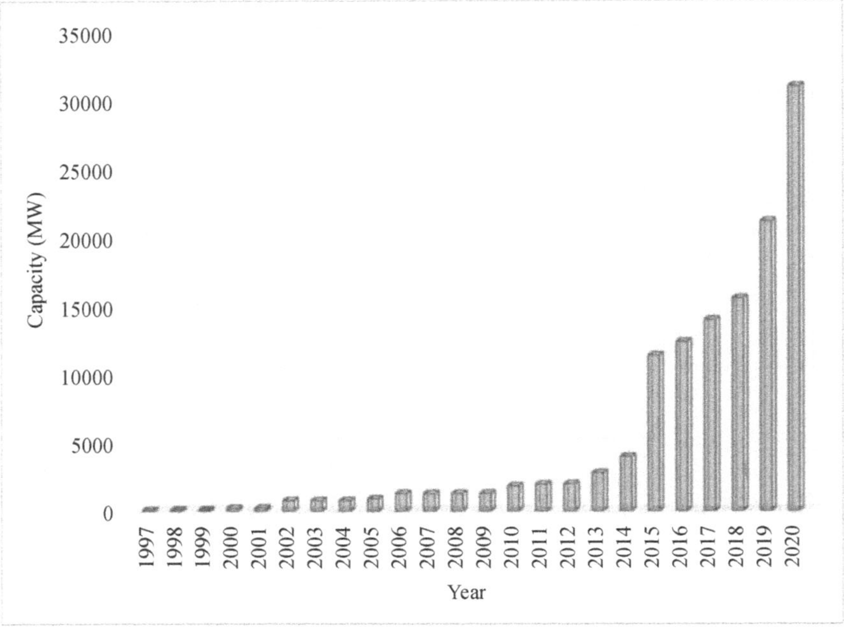

Fig. 1.1. Cumulative new capacity of VSC HVDC Transmission

Table 1.1: MMC project in operation across the world

	Project Name	**Country**	**Ratings**	**Year of commission**
1	Trans Bay Cable [37]	USA	400 MW, ±200 kV	2010
2	Inelfe [38]	France - Spain	2000MW, ±320 kV	2015
3	SylWin 1 [37]	Germany	900 MW, ±320 kV	2015
4	BorWin 3 [39]	Germany	900 MW, ±320 kV	2019
5	Cobra cable [40, 41]	Denmark - Netherlands	700 MW, ±320 kV	2019
6	ALEGrO [42, 43]	Belgium - Germany	1000 MW, ±400 kV	2020
7	elecLink [44]	UK - France	1000 MW, ±320 kV	2022

technical advances in voltage source converter (VSC) HVDC, specially with MMC-based alternatives. Fig. 1.1 shows the cumulative growth of VSC HVDC. The total installed capacity is now more than 30 GW.

This growth can be attributed, in particular, to the installation of MMC HVDC networks since 2014. MMC has facilitated the deployment of VSC HVDC with

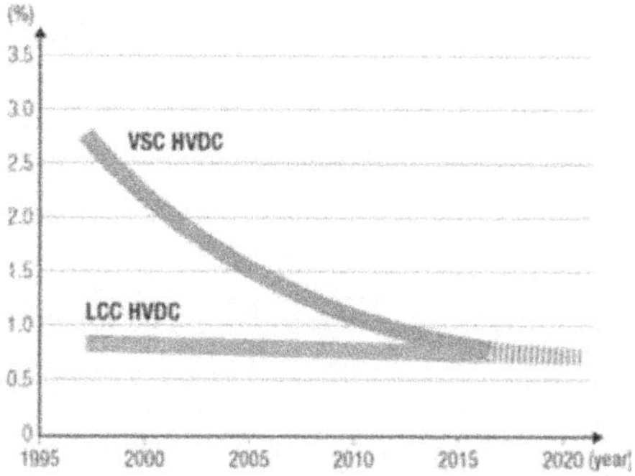

Fig. 1.2. Trend in electrical losses at HVDC converter station

capacities ranging from a few hundred MW to GW. Table 1.1 lists few major MMC projects currently in operation across the world [37, 45]. These projects use half-bridge submodules and are two-terminal HVDC systems. Two GW ULTRANET HVDC project being built in Germany, on the other hand, is the first to use a full-bridge submodule and the first multi-terminal MMC HVDC. The reduction of electrical losses at the converter station is another reason driving the growth of HVDC. Fig. 1.2 compares the converter station losses of a line commutated converter (LCC) with a voltage source converter (VSC). Losses have been reduced, making both technologies economically equal.

1.2.1 Indian scenario

The scenario in India is comparable to the global trend. In order to meet its ever-increasing energy demand, India is focusing on renewable energy sources as part of its commitment to the environment and sustainable development. HVDC is a preferred method for integrating various renewable energy sources. These are electric superhighways used to extract energy from remote and offshore places. Fig. 1.3 presents the Indian HVDC network. The technical details of these HVDC transmission systems are listed in Table 1.2 [46].

The Rihand-Dadri bipole, with a capacity of 1500 MW, is India's first HVDC transmission line, having been commissioned in 1991. In addition, India commis-

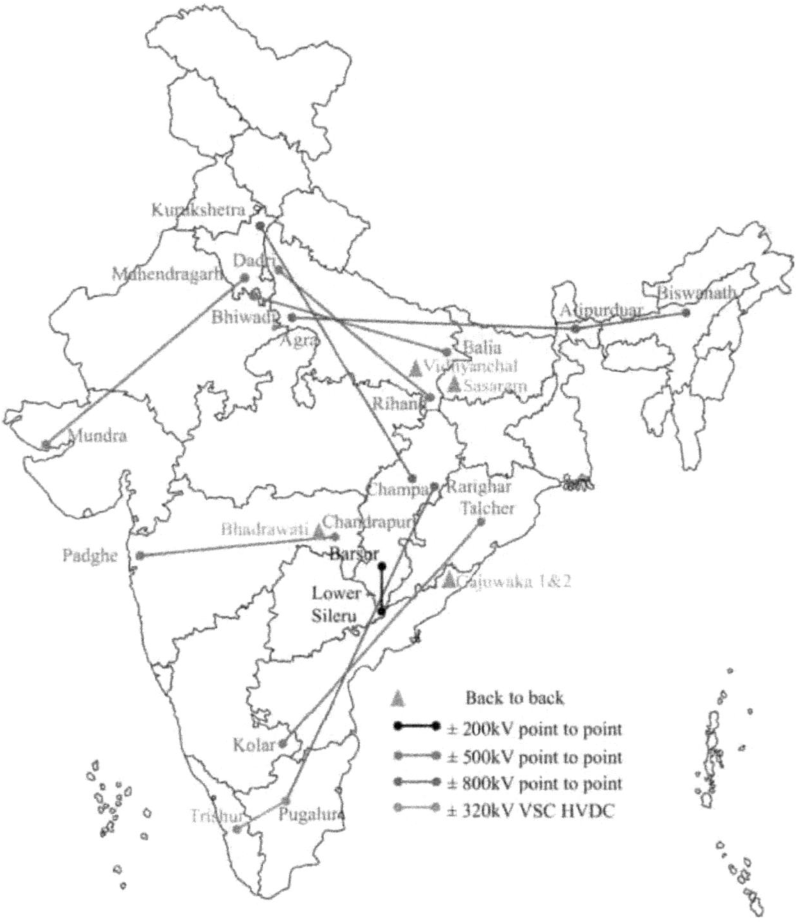

Fig. 1.3. Indian HVDC network

sioned its first VSC HVDC transmission system in 2021 [47]. This 2 GW transmission line in India's Pugalur and Trichur districts is rated at ±320 kV. There also exist cross-border links with Bangladesh and Bhutan. Several other HVDC projects are in the planning stages, including cross-border connections with Nepal, Bangladesh, Sri Lanka, and Myanmar [48]. The Maharashtra state of India is constructing the country's second VSC HVDC. In 2019, the Indian HVDC transmission system market was estimated to be worth $2 billion. It is expected to double in size by 2027 [49].

1.3 Modular Multilevel Converters

An MMC is primarily an arrangement of the submodules (also known as cells). The cascaded connection of these submodules forms a string known as an arm. Three similar arms can be connected to form a star configuration, as shown in Fig. 1.4a, or a delta structure, as shown in Fig. 1.4b [50, 51]. The single star and single delta configurations are three-terminal circuits (excluding neutral). When two single-star

Table 1.2: Indian HVDC projects

	Project Name	Voltage / Type	Capacity	Year of commission
1	Vindhyanchal BTB*	BTB	500 MW	1989
2	Basur-Lower Sileru	±200 kV	200 MW	1989
3	Rihand-Dadri	±500 kV	1500 MW	1991
4	Bhadrawati BTB	BTB	1000 MW	1997
5	Chandrapur-Padghe	±500 kV	1500 MW	1999
6	Gajuwaka 1 BTB	BTB	500 MW	1999
7	Sasaram BTB	BTB	500 MW	2002
8	Talcher-Kolar	±500 kV	2500 MW	2003**
9	Gajuwaka 2 BTB	BTB	500 MW	2005
10	Balia-Bhiwadi	±500 kV	2500 MW	2010
11	Mundra-Mahendra	±500 kV	2500 MW	2012
12	Agra-Biswanath	±800 kV	6000 MW	2017
13	Champa-kurukshetra	±800 kV	6000 MW	2017
14	Raigarh-Pugalur	±800 kV	6000 MW	2019
15	Pugalur-Trichur	±320 kV	2000 MW	2021

*BTB: Back to back **upgraded in 2007

cells are combined, they form the double-star configuration known as MMC, as shown in Fig. 1.5. It is a five-terminal circuit. Due to the lack of common dc-link terminals in both three-terminal circuits, they are ineligible for ac-dc or dc-ac power conversion. As a result, these are suitable for energy storage systems and STATCOM but unsuitable for HVDC or industrial motor drives in the presence of a common dc supply [52, 53]. The five-terminal circuits can access a common dc source through both neutral and are suitable for HVDC or motor drive.

1.3.1 Converter family and classification

Fig. 1.6 shows a classification of the converter family. MMC is the most recent and advanced member of this family. MMC is the most latest variant of the voltage source converter. Although MMC is a generic term for modular multilevel cascaded converters (MMCC), the most common MMCC topology is a double star chopper cell (DSCC) configuration [54]. Nowadays, MMCC-DSCC is directly referred to as MMC [55].

Different from MMCC-DSCC, MMCC is classified into various other typologies [51, 54–57]. These members differ in terms of the circuit configuration and SM types. They can be single, double, or triple star arrangements. The submodules

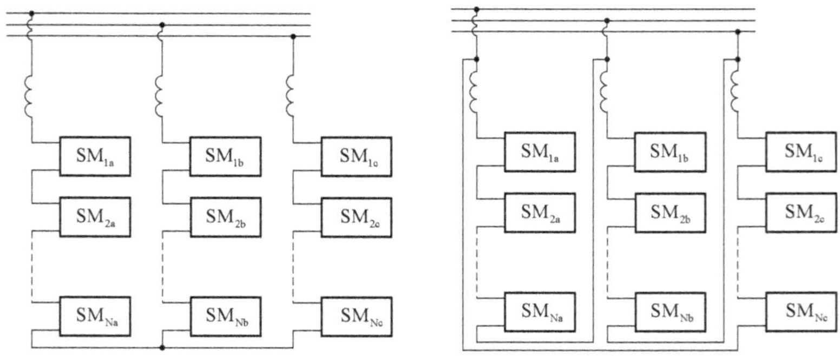

(a) (b)

Fig. 1.4. (a) Star configured MMC, (b) Delta configured MMC

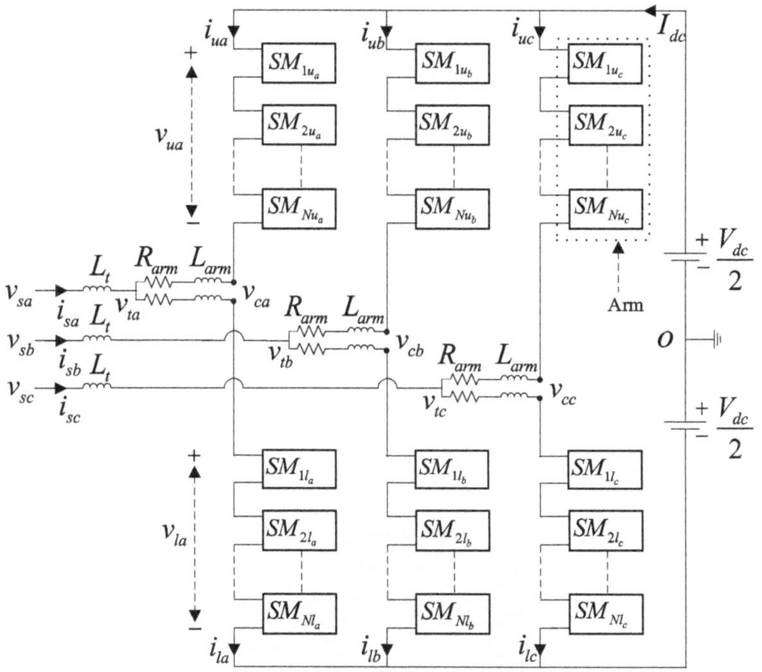

Fig. 1.5. MMC architecture

might be full-bridge (also known as bridge-cell) or half-bridge (also known as chopper cell) [54]. When any other type of submodule, or even a combination of two different submodules, is utilised, the configuration is known as a hybrid MMC.

1.4 Review of MMC HVDC Transmission

In the last two decades since its inception, the MMC HVDC has been explored extensively. In this section, the works related to the proposed thesis have been critically reviewed and presented.

1.4.1 Modelling and stability analysis of MMC HVDC

Many researchers have contributed to the MMC's modelling and operation. In [58, 59], various types of EMT models have been built and compared. Among the

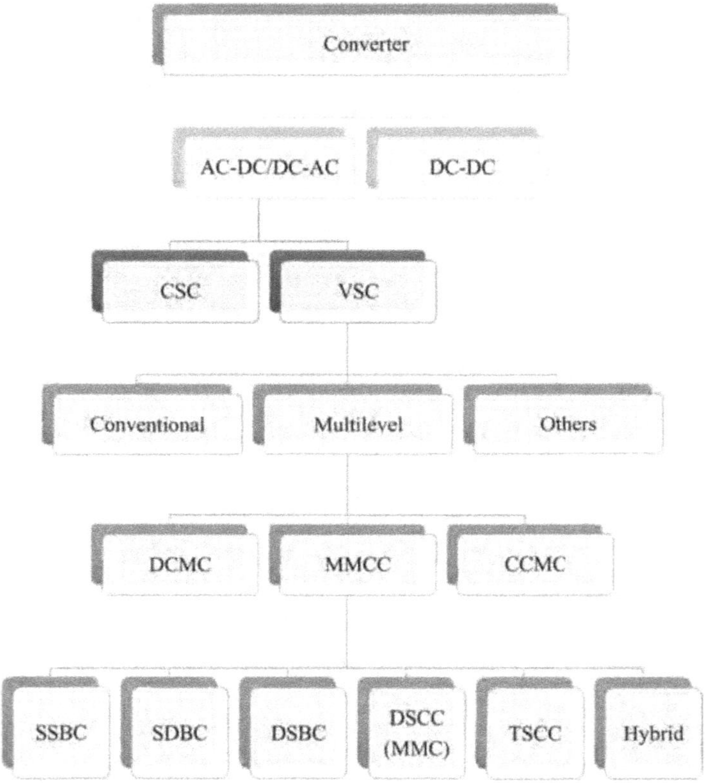

- AC-DC: ac to dc converter
- DC-AC: dc to ac converter
- DC-DC: dc to dc converter
- CSC: Current source converter
- VSC: Voltage source converter
- DCMC: Diode clamped multilevel converter
- CCMC: Capacitor clamped multilevel converter
- MMCC: Modular multilevel cascade converter
- SSBC: Single star bridge cell
- SDBC: Single delta bridge cell
- DSBC: Double star bridge cell
- DSCC: Double star chopper cell
- TSCC: Triple star chopper cell
- MMC: Modular multilevel converter

Fig. 1.6. Converter family

other approaches, the averaged value model (AVM) described in [58] is the most accurate and fast. However, it disregards redundancy and is inapplicable for submodule voltage balancing. It is unable to simulate the arm current, SM capacitor voltage, etc. The detailed equivalent model (DEM) presented in [59] is fast and accurate, similar to AVM, but requires a highly precise modulation technique to achieve minimum voltage deviation. In [60], authors have developed a detailed small-signal model of MMC. The modelling of 400 MW monopolar scheme of the Trans Bay Cable project, designed with 100 HBSMs in each arm, is presented by the authors in [61]. The wave function and multiscale modelling of the MMC have been investigated in a recent study [62]. This Fourier series approach models each submodule switching transient without compromising the accuracy. The authors of [63] investigate the admittance model for the MMC, and [64] describe an analytical model for faults in symmetrical monopoles. Because of the MMC's intricate structure, studies focus on specific aspects while neglecting others.

These models aid in establishing the relationship between the system's dynamics and its stability. Stability analysis of MMC has been investigated in [65, 66]. The authors of [65] perform a steady-state analysis and present a detailed insight into the interaction between currents and voltages. Nonetheless, they did not consider the MMC's behaviour in the presence of transients. The authors of [66] estimated the global asymptotic stability of MMC using the Lyapunov function. The authors of [60] investigated the harmonic instability of MMC and highlighted the prospect of the system being unstable due to the presence of internal dynamics. They suggested increasing arm resistance as a solution, however, this would lead to additional losses. The authors of [67] study the harmonic stability of MMC HVDC with passive circulating current filters. However, because circulating currents can be regulated via controllers, the usage of a passive filter is an inefficient strategy. The stability of the three-terminal MMC HVDC has been examined in [18].

1.4.2 Control approach of MMC

MMC control is a multifaceted challenge. The primary goals are power flow and voltage regulation. However, to assure efficiency and power quality, control objectives such as circulating current elimination, capacitor voltage balancing, voltage ripple minimization, and so on, should be met. The concept of MMC control is depicted in Fig. 1.7. In most cases, the fundamental current controller controls the power flow between ac and dc terminals, as well as the reactive power exchange between the converter and the ac grid. Circulating current controllers are installed in parallel with it. The arm switching functions are generated based on control commands from both. The arm switching function and voltage balancing command are used to form the final submodule control signal.

Various approaches have evolved over the last decade to meet each of the control objectives. The authors of [68] proposed a circulating current suppression strategy based on the multi-input multi-output non-parametric model. While the suggested

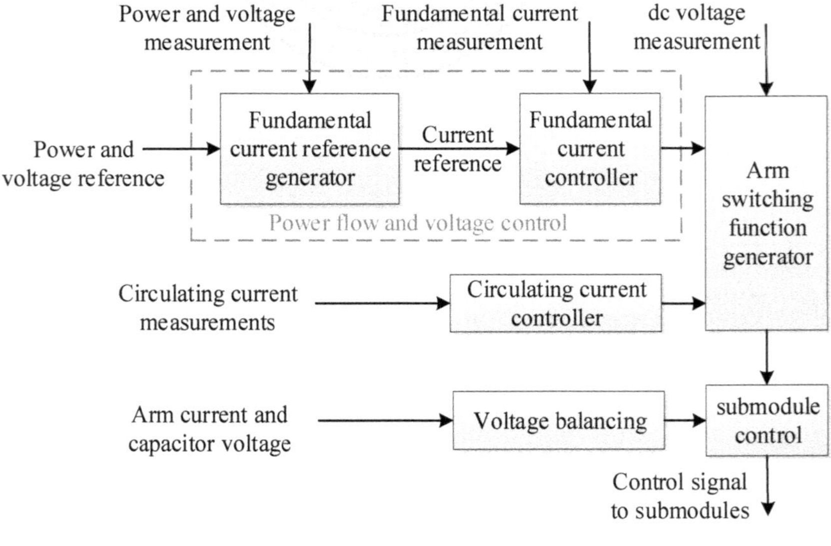

Fig. 1.7. Concept of MMC control

technique improves dynamic performance, it also increases the computational burden as determining controller values requires optimisation. A model predictive control is proposed in [69]. It lessens the calculation burden by decoupling SM capacitor voltage control from the cost function and preselecting control alternatives for the next sampling period. In [70], an indirect model predictive control is provided that defines the cost function using a mathematical model-based pre-derived stability margin. The drawbacks of predictive model control include a large computational burden, complexity, susceptibility to weighting factors, and variable switching frequency [71].

Arm current control is proposed by the authors of [72, 73]. In [72], arm-current measurements are used to determine phase currents and circulating currents, which are then translated into d-q axis variables. To achieve control objectives in the d-q coordinate, proportional-integral (PI) controllers are used. In [73], all six arm currents are directly controlled using a dedicated proportional-resonant (PR) controller for each arm. Although all arm currents can be controlled independently, this increases the number of controllers. Furthermore, the authors of [72], [73] do not investigate the interaction between active and reactive power. The authors of [74] derived six intermediate adjustable voltages. It resulted in decoupled control of each of the six arms. The total control structure, on the other hand, includes seven controllers, resulting in a very complex control architecture. Few researchers have also investigated energy-based control [75–77].

The internal dynamics of MMC are the most important aspect. The authors of [65] investigated how voltage and current interact within the SM. The internal dynamics of the MMC are a result of this interaction. The distribution of capacitors in six arms introduces complexities that must be addressed. However, most conventional techniques overlook the MMC's internal dynamics. The authors of [60] investigated the internal dynamics of the SM capacitor and its impact on stability. The authors of [78] explored the dynamics of MMC on an unbalanced grid. However, when designing the controllers, they adopted the standard PI-based controller, which ignores internal dynamics.

The authors of [69, 70] presented model predictive control taking the internal dynamics into account. The authors of [79] investigate the coupling of electrical quantities in SM while accounting for internal dynamics. They did not, however, account for 3rd order harmonic voltage ripple in the SM capacitor. They also neglected the internal dynamic for the auxiliary controller and applied the conventional approach.

1.4.3 Scope of transformerless MMC HVDC transmission

One of the advantages of the MMC is transformerless operation. However, it has not received much attention. Since transformers are very expensive and contribute significantly to converter station costs, a transformerless operation offers significant economic savings. Internal dynamics shown in [60, 65] and many other existing pieces of literature demonstrate that a third-order harmonic voltage can generate a third-order harmonic current. However, most studies have taken into account either an interface star-delta transformer or a virtually grounded dc terminal. It restricts the flow of third-order harmonic current. As a result, the third harmonic current has no detrimental effects on the system. This third-order harmonic current may have a severe impact on the system while operating in transformerless mode with grounded dc.

The authors in [80] provide the control of transformerless MMC HVDC during asymmetric grid disturbances. The authors of [81] developed a transformerless back-to-back system for connecting two Japanese radial feeders using MMC. However, both studies took into account a virtually grounded dc terminal. However, in an ungrounded MMC HVDC system, an imbalanced voltage between the two poles of the dc terminal occurs frequently and persists [82]. To ensure safe operation, zero potential should be provided, which can be achieved through artificial grounding. A physical ground connection provides a path for the third-order harmonic current to flow. The authors, on the other hand, have taken into account the interface transformer, which blocks this third-order current.

The authors of [83] developed a three-phase four-wire system that allows third harmonic currents to flow. An analytical controller has been proposed by the authors to suppress third-order harmonics current. However, the controller is reliant on SM capacitors and becomes less effective as the capacitors age. Furthermore, not all capacitors are identical in practice. There could be a variety of additional circumstances in which ac and dc terminals are connected to the ground, forming the conduction path for the third-order harmonic current.

1.4.4 Scope of multipolar MMC HVDC transmission

Another aspect of the MMC application is multipolar operation. The basic concept for multi-pole configuration is present in the available literature. However, a thorough investigation is necessary. Using the multipolar arrangement, many existing high voltage alternating current transmission lines can be converted to HVDC networks. It is possible to accomplish this by installing MMCs at both the sending and receiving ends. Existing healthy alternating current transmission wires can be used for HVDC transmission, reducing line investment. Since most modern alternating current transmission lines are three-phase, a three-wire system would be more appropriate for upgrading them to an HVDC transmission line. The authors of [84] presented a tri-pole employing a combination of HBSM-based MMC (H-MMC) and FBSM-based MMC (F-MMC). In addition, they developed a synchronised control technique for the same. However, the full potential of F-MMC is not being realised. The F-MMC-based pole serves as a backup by carrying the difference current between the other two poles, despite the fact that F-MMC is more expensive than H-MMC. The differential current would be zero if both H-MMC poles handled equal power. The authors of [85] proposed several symmetrical and unsymmetrical multipole MMC HVDC. However, the effect of one MMC on other under has not been considered by either one. The HBSM-based MMC suffers from the lack of dc fault handling capability.

1.4.5 Summary

- Researchers have proposed a variety of models of MMC. Due to a large number of components, however, these models sacrifice either efficiency or detail. Moreover, the intricate operation of MMC may result in harmonic instability.

- Multiple control strategies have been developed to address the complex control challenge posed by the MMC. However, the MMC's internal dynamics have been neglected in these approaches.

- The MMC provides the option for transformerless operation. It can significantly contribute to converter station cost savings. The possibility and associated complexities of transformerless operation are not yet investigated.

- In order to convert a three-phase, three-wire ac transmission line into an HVDC transmission line, an MMC-based multi-pole HVDC can be advantageous. The initial idea is documented in the literature. However, it requires a thorough investigation. Additionally, HBSM-based MMC is incapable of handling dc faults.

1.5 Motivation

HVDC makes a significant contribution to the rapidly expanding electric grid. The modular multilevel converter is the most modern and cutting-edge voltage source converter for VSC HVDC. MMC technology has evolved remarkably in the recent decade. It is an excellent alternative for high voltage and high power applications due to a number of advantages. The most cost-effective MMC topology is HBSM-based MMC, which lacks dc-fault handling capability. Topologies with fault-handling capabilities are complicated and expensive. Furthermore, the MMC has a slew of untapped advantages that have the potential to transform the transmission network. Summary for literature review and motivation of the proposed work are as follows:

- A diverse set of submodules has been developed, each with its own set of advantages and disadvantages. It is crucial to choose the appropriate one based on the application.

- The MMC operation causes the even circulating current to flow between the MMC legs. It is essential to minimise this without compromising the system's efficiency and stability.

- Despite the fact that the MMC is a complex structure, the controller design approaches are based on traditional converter models. It results in the coupling of active and reactive power. The models and the controller should be designed considering the complex structure to provide better decoupling.

- The MMC offers the opportunity to remove the transformer from the network. However, this may result in the flow of third-order zero-sequence current. It has a detrimental impact on efficiency and operational range. With sufficient analysis and a suitable control algorithm, the unwanted current can be eliminated, paving the way for transformerless operation.

- Another untapped application of the MMC is multipolar HVDC. While it may provide several benefits, its operation and associated complications are quite limited.

In view of the above, the following objectives are identified for the present thesis work.

- A comprehensive review of submodules to select an accurate MMC topology for the HVDC transmission.

- Development of EMT and small-signal model of MMC for stability analysis and estimation of the stable range of operation. An investigation of the circulating current controller requirement and its impact on the stability and internal dynamics has been carried out.

- To develop an internal dynamics-based detailed model for the fundamental and circulating currents, and to propose new controllers for the efficient operation of MMC and active-reactive power decoupling.

- To analyse the transformerless operation, its impact on efficiency, arm energy requirement and operating range and to update the dynamic equations of MMC. Development of a controller for third-order zero-sequence current elimination while ensuring stability and efficiency has also been carried out.

- To design and analyse symmetrical and asymmetrical multipolar MMC HVDC transmission systems and to derive the corresponding mathematical relationships. Fault analysis of the symmetrical tri-pole MMC HVDC network, as well as impact analysis following the change in the operating conditions of one MMC, is investigated and the control algorithm for the dc-fault ride-through capabilities of HBSM-based tri-pole MMC HVDC transmission has been proposed.

1.6 Thesis Organisation

To summarize the overall thesis work, the thesis has been divided into seven chapters. The initial phase of work is done on exploring the MMC architecture and operation. The second phase focuses on developing the controller for efficient operation, followed by the transformerless operation and multipolar operation with dc fault handling capability.

This chapter presents an overview of MMC and its contribution to the growth of VSC HVDC across the globe. A detailed literature report on the modelling, control, transformerless operation, and multipolar operation is included.

In **Chapter 2**, a comprehensive review of the MMC topologies and its submodules is presented to provide an overview of evolving technology. The MMC architecture and operation have been briefly elaborated. MMC operation is essentially an efficient control of the number of SMs inserted in each arm. The literature describes sixty different types of submodules. This chapter focuses on a few selected SMs. The switching signal's relationship with capacitor current and submodule voltage also has been included.

The MMC legs contain a circulating current. A dedicated controller is required for efficient operation. In **Chapter 3** an electromagnetic-transient (EMT) model of the MMC is built using a real-time digital simulator (RTDS) to analyse the circulating current's significance and impact. In addition, an equivalent small-signal model has been developed in MATLAB. The EMT model has been validated against the small-signal model. The stability analysis, including the eigenvalue and root-locus, has been performed. The investigation of stability limits with variation in controller variables has been carried out for a wide range. A study of the requirement and impact of the circulating current controller on the dynamics of the MMC is also presented. The eigenvalue shows that the presence of a circulating current controller enhances the stability of the system.

Given that MMC is a converter topology, the most typical controller design technique is to take the standard converter model and design the controller accordingly. This approach ignores the internal dynamics. However, MMC's modular architecture distributes the capacitors evenly among its six arms. This distribution generates a complex internal dynamic that significantly impacts the controller's operation and cannot be ignored. Taking into account the internal dynamics, a detailed model for the MMC's fundamental, and circulating currents have been developed in **Chapter 4**. The detailed modelling reveals cross-couplings and inter-couplings. The proposed controllers consider these couplings and their effect on the overall operation. These controllers provide superior power decoupling when compared to traditional control methods. The impact of inter-couplings, cross-couplings, and SMC voltage fluctuations on overall operation has been investigated. The comparative performance analysis of proposed controllers demonstrates that the proposed simplified controller outperforms the other control methods.

One of the MMC's advantages is its transformerless operation. However, the analysis in **Chapter 5** reveals that the internal dynamic of MMC induces a third-order zero-sequence harmonic current during this mode. Its effect on system operation, energy consumption, and power transmission capabilities have been investigated. The internal dynamic equations of the transformerless MMC configuration have been developed, and two distinct controllers (PR and PI) are presented to suppress the undesirable third-order current. The performance analysis of these controllers demonstrates that the third-order harmonic current is effectively suppressed. The electromagnetic transient models of MMC under different configurations have been developed on the real-time digital simulator. The state-space model has been developed, and the stability analysis has been performed using it. The effect of the controllers on the internal dynamics of the transformerless MMC is also studied.

MMC can play a significant role in upgrading ac transmission lines to dc transmission lines. However, it warrants further investigation. Muti-pole MMC HVDC is one of the few applications that remains mostly untapped. In **Chapter 6**, three-wire symmetrical and asymmetrical MMC HVDC configurations using two MMCs are proposed. The symmetrical configuration's general structure has been thoroughly examined. Both terminals are built with MMC that is based on a half-bridge, which lack the dc-fault handling capability. Mathematical relationships between pole variables are derived. Additionally, a proposed controller and control algorithm incorporates dc fault ride-through capability. The effectiveness has been demonstrated on a tri-pole.

Chapter 7 summarises the main findings of the work presented in this thesis and suggests a direction for future work in the field of MMC HVDC transmission.

Chapter 2

A Review of Modular Multi-level Converter (MMC)

2.1 Introduction

The history of MMC can be traced back to the mid-1970s [11, 12]. However, the MMC in the present version was proposed by Marquardt in the German patent [7] in 2002. Over a period of time, MMC has found its application in several areas such as high voltage dc (HVDC) [86, 87], medium voltage drives [88, 89], variable speed application [90], flexible ac transmission systems and static compensator [52, 91–93], power quality [94], dynamic braking of choppers [95], photovoltaic applications [96], wind energy [97], etc. The structural architecture of MMC offers several technical and economic advantages [98]. Modular structure establishes its dominance over other converters regarding operating cost and life cycle cost maintenance. The significant advantage of MMC is its ability to realize the transformerless connection between medium voltage system and grid [99]. The primary building block of an MMC is the submodule (SM), which facilitates the characteristic like scalable voltage level leading to staircase output waveforms, modularity, reduced harmonics and electromagnetic losses, high power quality, [25, 100, 101], etc.

The first practical application of MMC based system is developed at the Trans

Bay cable HVDC project by Siemens. It incorporates 216 half-bridge submodules in each arm, and the rated voltage of each SM capacitor is 1 kV, which makes it efficient to generate 201 output voltage levels [59,61,102]. INELFE is another initial MMC-HVDC-based link between France and Spain, having 401 voltage levels and the potential to transfer rated 2x1000 MW power [50, 102].

This chapter provides a comprehensive review of MMC topology and its submodules (SMs). The MMC architecture and its operation have been explained in brief. It is essentially the process of controlling the number of SMs inserted in each arm. There are over sixty different types of submodules described in the literature. This chapter discusses a few selected SMs. There is also an overview of the important concerns and challenges.

2.2 MMC Architecture

The MMC's basic architecture is depicted in Fig. 2.1. It is mostly made up of submodules (SMs) arranged in six identical arms. The N number of SMs, connected in series with an inductor L_{arm} and a resistance R_{arm}, forms an arm [100, 102]. A leg is formed by connecting two such arms in series. The ac terminal of the leg is where these two arms meet. Three legs, each dedicated to one phase of the three-phase ac system, are put together in parallel to make an MMC [58]. The ac grid is connected to MMC through interface transformer/reactor of inductance L_t. The junction points of the parallel legs act as upper and lower dc poles of the MMC and are connected to the dc source/network of the voltage V_{dc}. The dc source is divided into two equal halves, each with voltage $V_{dc}/2$. The mid-point 'o' can be physically or virtually grounded.

The fundamental frequency voltage at the midpoint of the MMC leg is denoted by $v_{tj,(j=abc)}$. The phase voltage of the ac grid is represented by $v_{sj,(j=abc)}$. $i_{sj(j=a,b,c)}$ is the current flowing through the interface transformer/inductor. Under balanced grid condition both v_{sj} and i_{sj} are of fundamental frequency. I_{dc} represents the dc current. Upper and lower arm currents are depicted by i_{uj} and i_{lj}, respectively.

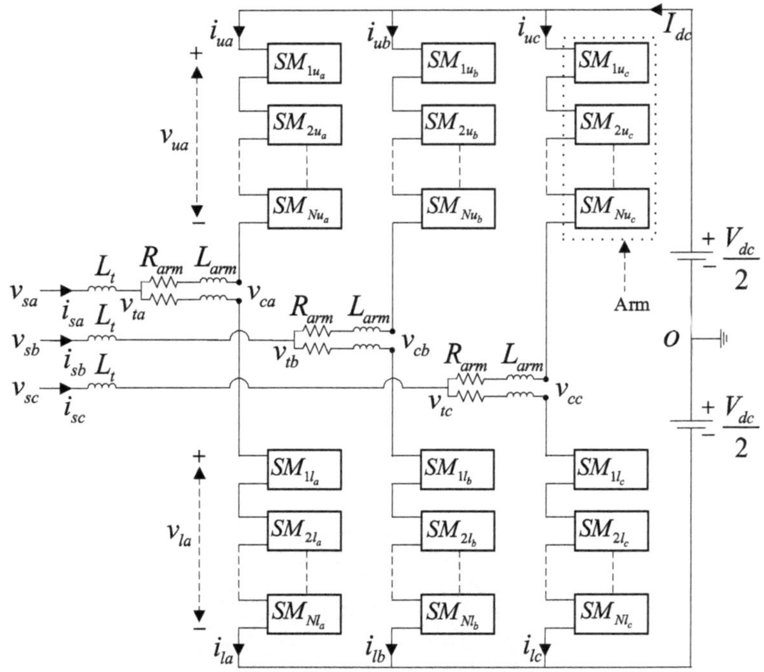

Fig. 2.1. MMC architecture

Similarly the upper and lower arm voltages are depicted by v_{uj} and v_{lj}, respectively.

2.3 MMC Operation

A submodule (SM) can perform three modes of operations: (i) inserted, (ii) bypassed, and (iii) blocked. When an SM is inserted, it permits current to flow through the capacitor and to charge/discharge it. In bypassed mode, switches of SMs operate in such a way that it allows current flow through SM but bypasses the capacitor. The capacitor remains charged. In block mode of operation, all the switches of SM are off, and current can flow only through anti-parallel diodes if possible [103]. Any particular submodule can be inserted, bypassed, or blocked based on switching commands [104]. Thus, any number of SMs can be inserted at a time corresponding to the switching scheme [105], and therefore, with appropriate switching schemes, an alternating voltage can be generated [106].

Each arm functions as a filter, absorbing power imbalance between the dc and ac sides. A high number of levels facilitate low switching frequency. Both lead to fewer losses, reduced harmonic distortion and smoother sine wave voltage, realising an almost filter-less structure [61, 107]. The ac output voltage of MMC can be seen as multilevel voltage synthesis. The waveform gets finer and finer with the higher number of SMs available in the arm structure to synthesise the waveform [101]. These advantages avoid the dc filter requirement and drastically reduce ac filter size, which can also be omitted via increasing the SM number in arms beyond 200 [100, 108, 109]. It should be noted that the number of voltage levels achieved in each phase is a function of both the modulation scheme and the switching function.

The arm voltage of a phase can be obtained as follows [60]:

$$v_u = N * S_u * u_{cu} \quad \text{and} \quad v_l = N * S_l * u_{cl} \tag{2.1}$$

where v_u and v_l are upper and lower arm voltages, respectively; N is the number of SMs per arm; S_u and S_l are the switching function for upper and lower arms, respectively, and u_{cu} and u_{cl} are the individual SM capacitor voltage in upper and lower arms, respectively.

Further, the current in each arm of MMC can be defined as: [60, 65]

$$\begin{aligned} i_u &= \frac{1}{3}I_{dc} - \frac{1}{2}I_s \sin(wt + \alpha_s) + I_{cir}\sin(2wt + \alpha_c) \\ i_l &= \frac{1}{3}I_{dc} + \frac{1}{2}I_s \sin(wt + \alpha_s) + I_{cir}\sin(2wt + \alpha_c) \end{aligned} \tag{2.2}$$

where I_{dc} is dc current which gets divided in three equal parts and flows in each leg. I_s and α_s are amplitude and phase, respectively, of phase current $i_{sj(j=a,b,c)}$. I_{cir} and α_c are the amplitude and phase angle of circulating current and can be defined as in (2.3) and (2.4), by [65]:

$$I_{cir} = \frac{\sqrt{(A\cos\alpha_s + B)^2 + (A\sin\alpha_s)^2}}{1 - \frac{N}{16\omega^2 C_{SM} L_{arm}} - \frac{M^2 N}{24\omega^2 C_{SM} L_{arm}}} \tag{2.3}$$

$$\alpha_c = \arctan(A\cos\alpha_s + B - A\sin\alpha_s) \tag{2.4}$$

where A and B are defined as:

$$A = \frac{3}{64}\frac{MNI_s}{\omega^2 C_{SM} L_{arm}} \qquad B = -\frac{1}{48}\frac{M^2 N I_{dc}}{\omega^2 C_{SM} L_{arm}} \tag{2.5}$$

where M is modulation index; N is number of SMs per arm; ω is fundamental angular frequency; C_{SM} is capacitance of SM capacitor and L_{arm} is arm inductance.

It is to be noted that the SM capacitor voltage ripple and MMC components rating have a compelling impact on circulating current [109]. Although the capacitor voltage of the SM capacitor would vary following charging and discharging, it must be maintained under the rated voltage of the capacitor [110, 111].

2.3.1 Voltage synthesis using submodules

MMC's operation is basically a voltage synthesis. Based on the switching signal, each submodule (SM) generates a voltage across its terminal. The arm voltage is obtained by adding the voltages of all submodules. To better comprehend the process, the author in [65] have explained the voltage synthesis considering the half-bridge depicted in Fig. 2.2. It is the ith SM of an arm. The i_x represents the current flowing through the SM. The SM capacitor voltage is denoted by u_{ci}. The capacitor is initially fully charged and has no ripples. It only has a dc voltage ($\overline{u}_c$) in this scenario. When the switch S_1 is turned on, and S_2 is turned off, the current i_x flows through the capacitor. Similarly, when S_1 is turned off and S_2 is turned on, the

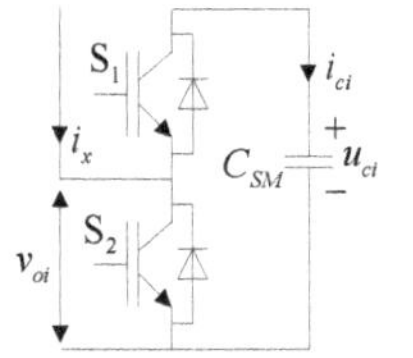

Fig. 2.2. Current and voltage in half-bridge submodule

current i_x does not pass through the capacitor. As a result, the current i_{ci} can be defined as follows [65]:

$$i_{ci} = S_i i_x \tag{2.6}$$

where S_i is the switching function of the SM. It is either '0'or '1'.

This current causes ripples in the capacitor voltage. The nth order frequency ripples in the capacitor voltage are defined as:

$$\Delta u_{ci}^{(n)} = \frac{i_{ci}^{(n)}}{jn\omega C_{SM}} \tag{2.7}$$

The capacitor voltage is defined as:

$$u_{ci} = \overline{u}_c + \sum_{n=1}^{\infty} \Delta u_{ci}^{(n)} \tag{2.8}$$

According to the switching signal, this voltage gets reflected SM terminal as:

$$v_{oi} = S_i \overline{u}_c + S_i \sum_{n=1}^{\infty} \Delta u_{ci}^{(n)} \tag{2.9}$$

Adding the terminal voltage of all the SMs of an arm we get the arm voltages for the upper and lower arm:

$$\begin{aligned} v_u &= \sum_{i=1}^{N} S_{ui} \overline{u}_c + \sum_{i=1}^{N} S_{ui} \sum_{n=1}^{\infty} \Delta u_{uci}^{(n)} \\ v_l &= \sum_{i=1}^{N} S_{li} \overline{u}_c + \sum_{i=1}^{N} S_{ui} \sum_{n=1}^{\infty} \Delta u_{lci}^{(n)} \end{aligned} \tag{2.10}$$

The subscript 'u' and 'l' denote the upper and lower arm variables, respectively.

Adding the upper and lower arm voltages gives the phase leg voltage. Likewise, subtracting the upper arm voltage from the lower arm voltage yield the voltage at the ac terminal of the leg.

2.4 Submodules (SMs)

Submodules are made up of transistors, diodes, and capacitors. The most common topologies of SMs are half-bridge and full-bridge SMs [93]. Numbers of scholars have proposed multiple multilevel modules for a variety of reasons, including dimension reduction, dc-fault handling capability, and increased voltage levels, although this complicates the structure and control [9]. Due to continued research to improve MMC's performance, the number of submodule versions has already surpassed 60 and is continuing to grow. This section explores a couple of these SMs in detail to facilitate comprehension. They are classified according to their output voltage levels. Table 2.1 compares a few selected submodules based on their architecture, voltage levels, and abilities to handle dc faults. To facilitate the discussion, SMs have been

Table 2.1: Comparison of different submodules

SMs	L	Voltage levels	S	M	BP	DFH
HBSM	2	$0, +u_c$	2	1	No	No
CSSM	2	$0, +u_c$	3	1	No	Yes
Unidirectional SM 1	2	$0, +u_c$	1	1	No	No
Unidirectional SM 2	2	$0, +u_c$	1	1	No	No
FBSM	3	$0, +u_c, -u_c$	4	2	yes	Yes
TLFCSM	3	$0, +u_c, +2u_c$	4	2	No	No
TLNPC 'i'	3	$0, +u_c, +2u_c$	4	2	No	No
TLNPC 'ii'	3	$0, +u_c, +2u_c$	6	2	No	No
TLNPC 'iii'	3	$0, +u_c, +2u_c$	4	2	No	No
DSM	3	$0, +u_c, +2u_c$	8	4	No	Yes
DBSM	3	$0, +u_c, -u_c$	2	1	Yes	Yes
TLCCSM	3	$0, +u_c, +2u_c$	5	3	No	Yes
IHSM	3	$0, +u_c, +2u_c$	5	3	No	Yes
CDSM 'i'	4	$0, +u_c, +2u_c, -u_c$	5	3	Yes	Yes
CDSM 'ii'	4	$0, +u_c, +2u_c, -u_c$	7	3	Yes	Yes
Asymmetrical	4	$0, +u_c, -u_c, -2u_c$	4	2	Yes	Yes
Mixed SM	4	$0, +u_c, -u_c , -2u_c$	6	3	Yes	Yes
SDSM	4	$0, +u_c, +2u_c, -u_c$	5	3	Yes	Yes
AFBCCDSM	4	$0, +2u_c, 4u_c, -2u_c$	6	3	Yes	Yes
FLCCSM	5	$0, +u_c, +2u_c, -u_c, -2u_c$	6	3	Yes	Yes
CCSM	5	$0, +u_c, +2u_c, -u_c, -2u_c$	8	4	Yes	Yes
PCSM	5	$0, +u_c, +2u_c, -u_c, -2u_c$	8	4	Yes	Yes

L – No of voltage levels; S – Total no of switches; M – Maximum number of switches in conduction path; BP – Bipolar operation; DFH – dc fault handling capability classified according to their output voltage levels. The remaining submodules are included in the appendix.

2.4.1 Two-level submodules

Half-bridge submodule (HBSM), as shown in Fig. 2.3a, consists of a capacitor and two IGBT switches. It is the simplest configuration capable of producing unipolar voltage by chopping the dc-link voltage [112]. It is sometimes referred to as a half-bridge chopper cell [54]. Additionally, it is known to as a 2L submodule [9], three-pole commutation cell [113], and bidirectional chopper [55]. As shown in Table 2.1, the HBSM is capable of producing two output voltages: 0 when bypassed and $+uc$ when inserted, using switches S_1 and S_2 [16, 55, 95], and hence, there is always a dc component in the MMC arm voltage generated by HBSMs [114]. The current and voltage of the HBSM can be defined as in (2.11). The current passing the SM and the capacitor are denoted by i_x and i_c, respectively. u_c is the capacitor voltage. S_{Si} is the switching signal for i_{th} switch of the SM.

$$\begin{aligned} i_c &= (S_{S1}\overline{S}_{S2}) i_x \\ v_o &= (S_{S1}\overline{S}_{S2}) u_c \end{aligned} \tag{2.11}$$

Due to the half-bridge module's capability of bidirectional current flow but unidirectional voltage blocking, its operation is limited to two voltage levels [113]. Although the presence of only two switches in the SM structure results in high efficiency for half-bridge SM-based MMCs [114], the presence of a dc component in the arm voltage restricts its applicability to MMCs coupled to dc systems only [94]. Additionally, a significant disadvantage of HBSM is its inability to handle dc fault current [95]. Handling dc-link short circuit fault current is a significant problem, as the anti-parallel diodes in the HBSM act as a rectifier-bridge when the IGBT switches are in off mode [65, 108]. The ac side continues to feed the fault current until the circuit breaker trips, at which point the system may take up to ten seconds

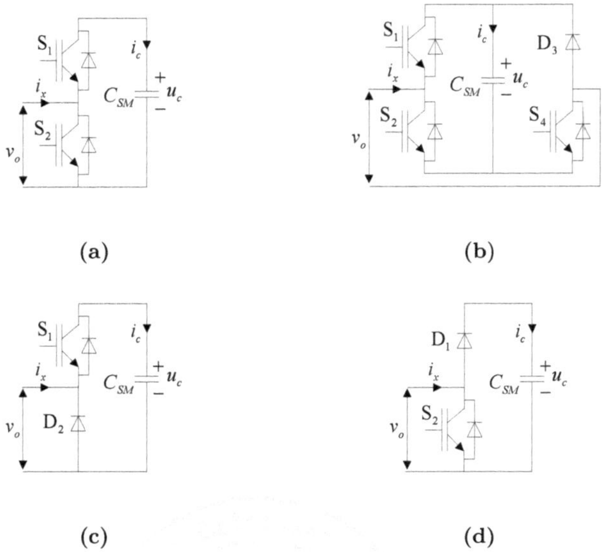

Fig. 2.3. Different two-level SM structures (a) HBSM, (b) CSSM (c) unidirectional SM: Type 1, (d) unidirectional SM: Type 2

to reset [115, 116].

Circuit configuration shown in Fig. 2.3b is a clamp single submodule (CSSM). This unipolar SM is a derived from full bridge SM (FBSM) configuration shown in Fig. 2.4a (discussed in section 2.4.2). The S_3 IGBT along with its antiparallel diode of the full bridge module is replaced by a diode D_3 to obtain improved configuration. The primary motive behind this modification was to get an SM with dc handling capability without significant sacrifice in terms of efficiency and cost. The current and voltage for the CSSM is defined in (2.12). Under the normal operating conditions, the unipolar voltage SM can operate in either inserted or bypassed mode through proper controlling of S_1 and S_2 switches, keeping S_4 permanently ON. The current and voltage of the SM get updated as in (2.13). Thus, the module is capable of generating 0 and $+u_c$ output voltage levels. Under a dc-link short circuit fault condition, the fault current takes the path through antiparallel diode of S_1, capacitor and antiparallel diode of S_4 i.e. similar to the fault current path in FBSM and thus gets blocked due to the presence of capacitor [33, 114]. This makes unipolar voltage SM capable of handling dc-link short circuit fault. The significant drawback of this configuration is that the overall conduction losses of SM are more than the same for FBSM as under normal operating conditions.

$$\begin{aligned} i_c &= (S_{S1}\overline{S}_{S2}S_{S4} - \overline{S}_{S1}S_{S2}\overline{S}_{S4})i_x \\ v_o &= (S_{S1}\overline{S}_{S2}S_{S4})u_c \end{aligned} \tag{2.12}$$

$$\begin{aligned} i_c &= (S_{S1}\overline{S}_{S2})i_x \\ v_o &= (S_{S1}\overline{S}_{S2})u_c \end{aligned} \tag{2.13}$$

Two-level SMs are inherently uni-polar but the current flow can be either bi-directional or unidirectional. Two SMs of unidirectional current flow are shown in Figs. 2.3c and 2.3d. This is an approach to reducing the semiconductor components per SM. As clear in the figures, it is derived from HBSM, by replacing one of its switches with a diode. The SM observes a restriction over switching state due to its dependency on the current direction [94, 117]. The current and voltage for the Type 1 and Type 2 unidirectional SM is defined in (2.14) and (2.15), respectively. Multiple other unidirectional configurations are proposed by De-Sousa et al. [117]. Similar to HBSM, these SMs lacks the dc fault handling capability.

$$\begin{aligned} i_c &= (S_{S1})i_x \\ v_o &= (S_{S1})u_c \end{aligned} \tag{2.14}$$

$$\begin{aligned} i_c &= (\overline{S}_{S2})i_x \\ v_o &= (\overline{S}_{S2})u_c \end{aligned} \tag{2.15}$$

2.4.2 Three-level submodules

Fig. 2.4a depicts a full-bridge submodule (FBSM) [33]. It is made of four switches and a capacitor. The relationship of current and voltage with the switching signal is defined in (2.16). The proper operation of switches S_1 to S_4 can provide bipolar output voltage levels 0, $+u_c$, and $-u_c$ [55, 95]. Thus, the FBSM-based MMC can be connected to both dc and ac systems [94]. Further, it is capable of handling dc-link short circuit fault current via generating reverse voltage through capacitor voltage [95, 108]. FBSM-based MMC can be fully disconnected by switching off all of the switches, whereas the same was not possible with HBSM [55]. The major drawback of FBSM is that its number of semiconductor devices is twice of that HBSM, making it more costly and less efficient [94, 95, 108]. A double commutation cell (i.e. FBSM) can be obtained by parallel connection of two identical HBSMs, which can give all four-quadrant operation [113].

$$
\begin{aligned}
i_c &= (S_{S1}\overline{S}_{S2}\overline{S}_{S3}S_{S4} - \overline{S}_{S1}S_{S2}S_{S3}\overline{S}_{S4})i_x \\
v_o &= (S_{S1}\overline{S}_{S2}\overline{S}_{S3}S_{S4} - \overline{S}_{S1}S_{S2}S_{S3}\overline{S}_{S4})u_c
\end{aligned} \tag{2.16}
$$

Three-level flying capacitor submodule (TLFCSM) [94, 113, 114], shown in Fig. 2.4b, has four switches, equal to that of FBSM and two capacitors. The capacitor C_{SM1} operating voltage is twice of capacitor C_{SM2} [9, 118]. The currents and voltage is defined in (2.17). The major drawback of this configuration is the lack of dc-link short circuit current handling capability [95]. Considering its disadvantage, a high number of switches and capacitors, high cost and complex control, it is not an attractive configuration [9, 119].

$$
\begin{aligned}
i_{c1} =& (S_{S1}\overline{S}_{S2} + \overline{S}_{S1}S_{S2})i_x \\
i_{c2} =& (S_{S1}\overline{S}_{S2}\overline{S}_{S3}S_{S4} - \overline{S}_{S1}S_{S2}S_{S3}\overline{S}_{S4})i_x \\
v_o =& (S_{S1}\overline{S}_{S2}\overline{S}_{S3}S_{S4})u_c + (S_{S1}\overline{S}_{S2}S_{S3}\overline{S}_{S4})2u_c + (\overline{S}_{S1}S_{S2}S_{S3}\overline{S}_{S4})(-u_c + 2u_c)
\end{aligned} \tag{2.17}
$$

Dargahi et al. [120] have developed an improved double flying capacitor cell configuration to enhance voltage quality. Three different configurations of three-level neutral point clamped submodule (TLNPCSM) are shown in Figs. 2.4c, 2.4d

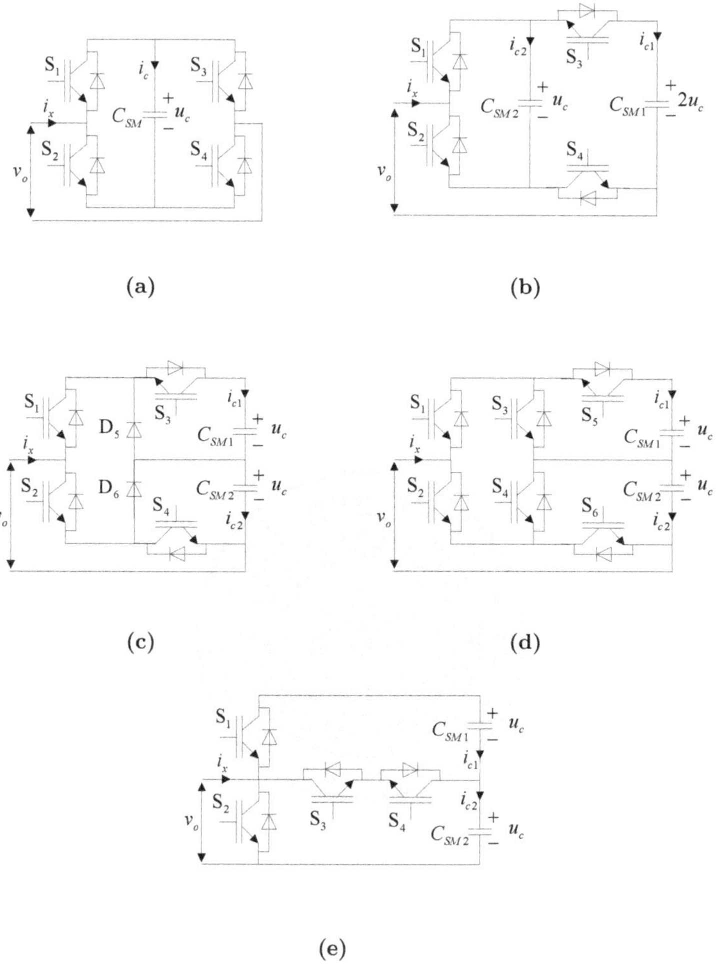

Fig. 2.4. Different three-level SM structures (a) FBSM, (b) TLFCSM (c) TLNPCSM structure 'i', (d), TLNPCSM structure 'ii', (e) TLNPCSM structure 'iii'

and 2.4e [9, 95, 113, 114]. The SM in Fig. 2.4c is almost similar to the TLFCSM in structure and generates three terminal voltages of 0, $+u_c$ and $+2u_c$ [9]. These voltages are derived from (2.18).

$$\begin{aligned} i_{c1} &=(S_{S1}\overline{S}_{S2}S_{S3}\overline{S}_{S4})i_x \\ i_{c2} &=(S_{S1}\overline{S}_{S2}S_{S3}+\overline{S}_{S1}S_{S2}\overline{S}_{S3})\overline{S}_{S4}i_x \\ v_o &=(S_{S1}\overline{S}_{S2}S_{S3}\overline{S}_{S4})2u_c+(\overline{S}_{S1}S_{S2}\overline{S}_{S3}\overline{S}_{S4})u_c \end{aligned} \tag{2.18}$$

Submodules in Figs. 2.4c and 2.4e contain four switches, two capacitors, and two diodes, while the submodule in Fig. 2.4d comprises six switches and two capacitors. The submodule in Fig. 2.4c can be obtained by connecting to a basic commutation cell in series [121]. In contrast, the submodule in Fig. 2.4e can be derived using the T-connection of semiconductor switches, and the midpoint switches must perform blocking in both polarities [122]. The currents and voltage of the structure 'ii' and 'iii', can be defined as in (2.19) and (2.20), respectively. The structure 'i' is uni-polar and bidirectional. However, the structures 'ii' and 'iii' are uni-polar as well as unidirectional. The incapability of handling dc-link short circuit fault current can be considered a major disadvantage of TLNPCSM [95, 114]. Furthermore, similar to TLFCSM, from a manufacturer's perspective; it is not an attractive configuration because of the high number of components, complex control and losses [9].

$$\begin{aligned} i_{c1} &=(S_{S1}\overline{S}_{S2}\overline{S}_{S3}\overline{S}_{S4}S_{S5}\overline{S}_{S6})i_x \\ i_{c2} &=(S_{S1}\overline{S}_{S2}\overline{S}_{S4}(\overline{S}_{S3}S_{S5}+S_{S3}\overline{S}_{S5})+\overline{S}_{S1}S_{S2}\overline{S}_{S3}S_{S4}\overline{S}_{S5})\overline{S}_{S6}i_x \\ v_o &=(S_{S1}\overline{S}_{S2}S_{S3}\overline{S}_{S4}+\overline{S}_{S1}S_{S2}\overline{S}_{S3}S_{S4})\overline{S}_{S5}\overline{S}_{S6}u_c \\ &\quad+(S_{S1}\overline{S}_{S2}\overline{S}_{S3}\overline{S}_{S4}S_{S5}\overline{S}_{S6})2u_c \end{aligned} \tag{2.19}$$

$$\begin{aligned} i_{c1} &=(S_{S1}\overline{S}_{S2}\overline{S}_{S3}\overline{S}_{S4})i_x \\ i_{c2} &=(S_{S1}\overline{S}_{S3}\overline{S}_{S4}+\overline{S}_{S1}S_{S3}S_{S4})\overline{S}_{S2}i_x \\ v_o &=(S_{S1}\overline{S}_{S3}\overline{S}_{S4})2u_c+(\overline{S}_{S1}S_{S3}S_{S4})\overline{S}_{S2}u_c \end{aligned} \tag{2.20}$$

A double submodule (DSM) configuration proposed by Ilves et al. [123], as shown in Fig. 2.5a, is derived by connecting two FBSMs to enhance the capacitor voltage balancing without increasing the power rating of SMs. The operation of the DSM can be defined as in (2.21). It yields three voltage levels as: 0, $+u_c$ and $+2u_c$ and holds the capability of dc fault ride through. Although, unlike FBSM, DSM is not bipolar, its maximum blocking voltage is twice of that FBSM.

$$
\begin{aligned}
i_{c1} =&\overline{S}_{S1}\overline{S}_{S3}(S_{S2}\overline{S}_{S4}(\overline{S}_{S5}S_{S6}\overline{S}_{S7}\overline{S}_{S8} + S_{S5}\overline{S}_{S6}S_{S7}S_{S8}) + \overline{S}_{S2}S_{S4}(\overline{S}_{S5}\overline{S}_{S6}\overline{S}_{S7}S_{S8} \\
&+ S_{S5}S_{S6}S_{S7}\overline{S}_{S8}))i_x \\
i_{c2} =&((S_{S1}\overline{S}_{S3}S_{S5}\overline{S}_{S7} + \overline{S}_{S1}S_{S3}\overline{S}_{S5}S_{S7})\overline{S}_{S2}\overline{S}_{S4} + (S_{S3}S_{S4} + \overline{S}_{S3}\overline{S}_{S4})\overline{S}_{S1}S_{S2}S_{S5}\overline{S}_{S7} \\
&+ (S_{S1}S_{S2} + \overline{S}_{S1}\overline{S}_{S2})\overline{S}_{S3}S_{S4}\overline{S}_{S5}S_{S7})\overline{S}_{S6}\overline{S}_{S8}i_x \\
v_o =&\overline{S}_{S6}\overline{S}_{S8}(\overline{S}_{S2}\overline{S}_{S4}(S_{S1}\overline{S}_{S3}S_{S5}\overline{S}_{S7} + \overline{S}_{S1}S_{S3}\overline{S}_{S5}S_{S7}) + S_{S2}S_{S4}(\overline{S}_{S1}S_{S3}S_{S5}\overline{S}_{S7} \\
&+ S_{S1}\overline{S}_{S3}\overline{S}_{S5}S_{S7}))u_c + \overline{S}_{S1}\overline{S}_{S3}\overline{S}_{S6}\overline{S}_{S8}(S_{S2}\overline{S}_{S4}S_{S5}\overline{S}_{S7} + \overline{S}_{S2}S_{S4}\overline{S}_{S5}S_{S7})2u_c
\end{aligned}
\tag{2.21}
$$

Three-level cross-connected SM (TLCCSM) is shown in Fig. 2.5c, proposed by Qin et al. [114]. It is primarily derived from the clamp double submodule (CDSM) and five-level cross-connected submodule (FLCCSM). The CDSM and FLCCSM are discussed in the following section. In TLCCSM, two half-bridge SMs are back-to-back cross-connected using a switch and a diode. Under the normal operation, the switch S_5 is permanently ON. The current and voltage for the same are defined in (2.22). With this, the derived module can give terminal voltage equal to 0, $+u_c$, and $2u_c$. During the fault condition, with all the switches in OFF condition, the fault current flows through the anti-parallel diodes and capacitors. With this, the TLCCSM gets the ability to block the fault current. Thus, similar to FBSM-based MMC, TLCCSM-based MMC also can generate reverse voltages V_{dc} in each arm, and both take equal time to reduce fault current to zero [114].

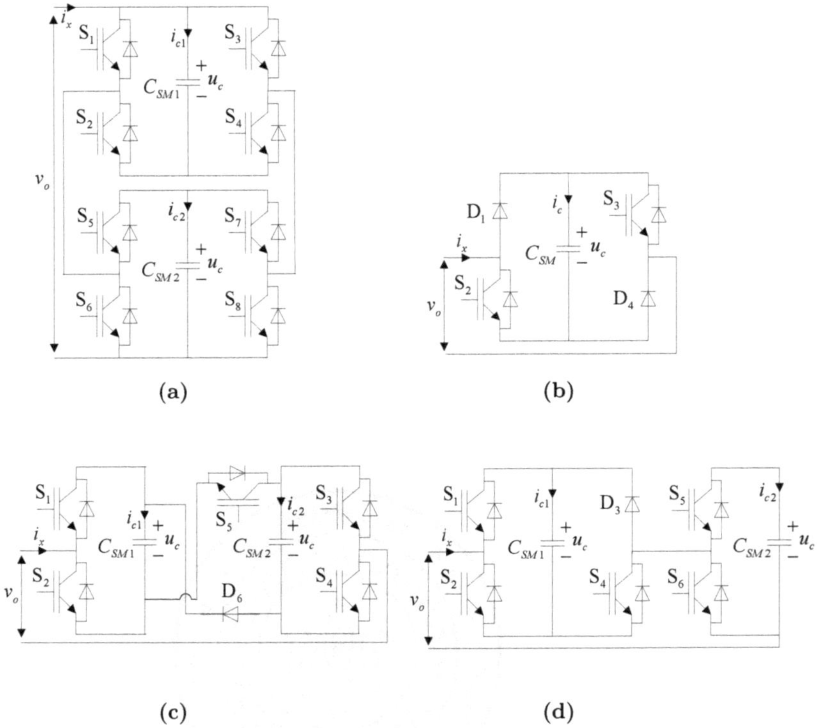

Fig. 2.5. Different three-level SM structures (a) DSM (b) DBSM (c) TLCCSM (d) IHSM.

$$\begin{aligned} i_{c1} &= S_{S1}\overline{S}_{S2}(S_{S3}\overline{S}_{S4} + \overline{S}_{S3}S_{S4})i_x \\ i_{c2} &= \overline{S}_{S3}S_{S4}(S_{S1}\overline{S}_{S2} + \overline{S}_{S1}S_{S2})i_x \\ v_o &= (S_{S1}\overline{S}_{S2}S_{S3}\overline{S}_{S4} + \overline{S}_{S1}S_{S2}\overline{S}_{S3}S_{S4})u_c + (S_{S1}\overline{S}_{S2}\overline{S}_{S3}S_{S4})2u_c \end{aligned} \tag{2.22}$$

Fig. 2.5b depicts a diagonal bridge submodule (DBSM). It is a modification of the full-bridge submodule by replacing two diagonal switches with diodes. The new configuration proposed by Yu et al. [124]. Based on (2.23) the SM gives three output voltages: $+u_c$, $-u_c$ and 0, similar to full-bridge SM. However, it has only two switches as in the half-bridge submodule. An improved hybrid submodule (IHSM) is derived from the clamp double submodule (Fig. 2.5d) through modification in connections [33]. It also can be seen as a combination of CSSM and HBSM. Two capacitors are being controlled using five switches and one diode. Connecting the two capacitors in series can produce a maximum blocking of $+2u_c$. Under the normal operation condition, the switch S_4 is permanently ON. Under this condition, the current and voltages of the IHSM are defined in (2.24).

$$\begin{aligned} i_c &= (\overline{S}_{S2}\overline{S}_{S3})i_x \\ v_o &= (\overline{S}_{S2}\overline{S}_{S3})u_c - (S_{S2}S_{S3})u_c; \end{aligned} \tag{2.23}$$

$$\begin{aligned} i_{c1} &= S_{S1}\overline{S}_{S2}(S_{S5}\overline{S}_{S6} + \overline{S}_{S5}S_{S6})i_x \\ i_{c2} &= S_{S5}\overline{S}_{S6}(S_{S1}\overline{S}_{S2} + \overline{S}_{S1}S_{S2})i_x \\ v_o &= (S_{S1}\overline{S}_{S2}\overline{S}_{S5}S_{S6} + \overline{S}_{S1}S_{S2}S_{S5}\overline{S}_{S6})u_c + (S_{S1}\overline{S}_{S2}S_{S5}\overline{S}_{S6})2u_c \end{aligned} \tag{2.24}$$

2.4.3 Four-level submodules

Figs. 2.6a and 2.6b depict two alternate configurations of clamp double submodule (CDSM) [60, 95, 114]. Both configurations combine two half-bridge submodules using additional switches and diodes [33, 125]. In Fig. 2.6a, switch S_5 is permanently ON, and the submodule operates as two HBSMs connected in series under normal operation conditions [95, 113]. The currents and voltage for the same are defined in (2.25). Both capacitors operate in parallel connection for CDSM, when the IGBT switch S_5 is blocked [126]. Further, the submodule can be operated as FBSM by parallel connection of both capacitors of different voltages. To avoid paralleling issues, diodes of the parallel path can be replaced using switches, as shown in Fig. 2.6b. Similar to the previous case, S_5 is permanently ON during the normal operation. The current and voltage for the same are defined in (2.26). The CDSM is capable of generating reverse voltage and thus, has dc-link current handling capability. It generates reverse voltage per arm equal to $\frac{1}{2}V_{dc}$ [127]. During a fault, with all switches in OFF condition, the fault current passes through the capacitors and

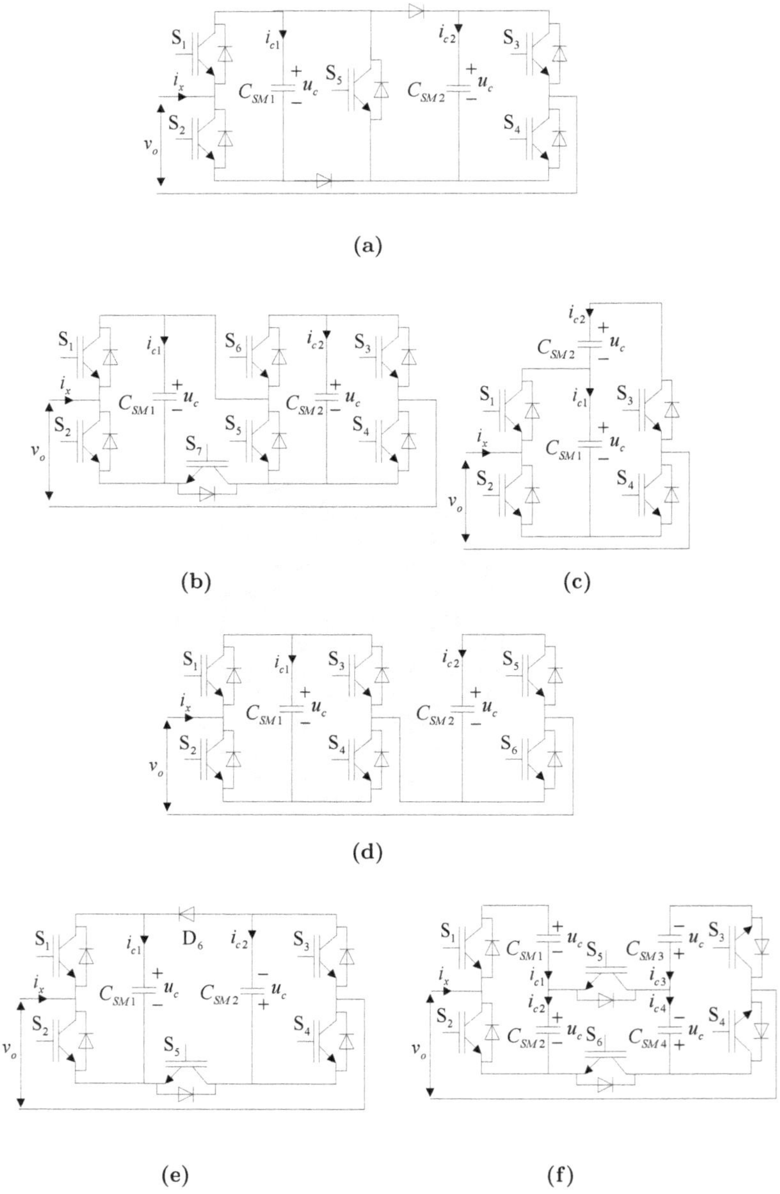

Fig. 2.6. Different four-level SM structures (a) CDSM structure 'i', (b) CDSM structure 'ii', (c) asymmetrical SM, (d) Mixed SM, (e) SDSM, (F) AFBCCDSM

anti-parallel diodes of the switches. Thus, the fault gets current is blocked. In terms of losses, considering the same voltage levels, CDSM stands in between HBSM and FBSM [108].

$$
\begin{aligned}
i_{c1} &=(S_{S3}\overline{S}_{S4} + \overline{S}_{S3}S_{S4})(S_{S1}\overline{S}_{S2} - \overline{S}_{S1}S_{S2})i_x \\
i_{c2} &=(S_{S1}\overline{S}_{S2} + \overline{S}_{S1}S_{S2})(\overline{S}_{S3}S_{S4} - S_{S3}\overline{S}_{S4})i_x \\
v_o &=(S_{S1}\overline{S}_{S2}\overline{S}_{S3}S_{S4} - \overline{S}_{S1}S_{S2}S_{S3}\overline{S}_{S4})u_c
\end{aligned} \tag{2.25}
$$

$$
\begin{aligned}
i_{c1} &=(S_{S3}\overline{S}_{S4} + \overline{S}_{S3}S_{S4})(S_{S1}\overline{S}_{S2}\overline{S}_{S6}S_{S7} - \overline{S}_{S1}S_{S2}S_{S6}\overline{S}_{S7})i_x \\
i_{c2} &=(S_{S1}\overline{S}_{S2} + \overline{S}_{S1}S_{S2})(\overline{S}_{S3}S_{S4}S_{S6}\overline{S}_{S7} - S_{S3}\overline{S}_{S4}\overline{S}_{S6}S_{S7})i_x \\
v_o &=(S_{S1}\overline{S}_{S2}\overline{S}_{S3}S_{S4} - \overline{S}_{S1}S_{S2}S_{S3}\overline{S}_{S4})(S_{S6}\overline{S}_{S7} + \overline{S}_{S6}S_{S7})u_c
\end{aligned} \tag{2.26}
$$

Asymmetrical SM, shown in Fig. 2.6c, is an alternative way of doubling the commutation cell. It got its name because the connected cells have two different voltage levels and can generate four voltage levels at the output terminal. Asymmetrical SM is capable of handling dc-link fault current. These voltages and the capacitor current can be obtained from (2.27). Fig. 2.6d shows a hybrid configuration obtained by combining HBSM and FBSM. Being a combination of two basic submodule structures, this SM can provide both unipolar and bipolar operations. Furthermore, the presence of FBSM furnishes it with dc handling capability, and the series connection of FBSM and HBSM facilitates the ability to generate four voltage levels [113]. The relationship of the currents and voltages with switching operation is defined in (2.28).

$$
\begin{aligned}
i_{c1} &=(S_{S1}\overline{S}_{S2}\overline{S}_{S3}S_{S4} - \overline{S}_{S1}S_{S2}S_{S3}\overline{S}_{S4})i_x \\
i_{c2} &=(-S_{S1}\overline{S}_{S2}S_{S3}\overline{S}_{S4} - \overline{S}_{S1}S_{S2}S_{S3}\overline{S}_{S4})i_x \\
v_o &=(S_{S1}\overline{S}_{S2}\overline{S}_{S3}S_{S4})u_c - (S_{S1}\overline{S}_{S2}S_{S3}\overline{S}_{S4})u_c - (\overline{S}_{S1}S_{S2}S_{S3}\overline{S}_{S4})2u_c
\end{aligned} \tag{2.27}
$$

$$
\begin{aligned}
i_{c1} =&(S_{S1}\overline{S}_{S2}\overline{S}_{S3}S_{S4} - \overline{S}_{S1}S_{S2}S_{S3}\overline{S}_{S4})(S_{S5}\overline{S}_{S6} + \overline{S}_{S5}S_{S6})i_x \\
i_{c2} =& - (S_{S1}\overline{S}_{S2} + \overline{S}_{S1}S_{S2})(S_{S3}\overline{S}_{S4} + \overline{S}_{S3}S_{S4})S_{S5}\overline{S}_{S6}i_x \\
v_o =&((S_{S1}\overline{S}_{S2}\overline{S}_{S3}S_{S4} - \overline{S}_{S1}S_{S2}S_{S3}\overline{S}_{S4})\overline{S}_{S5}S_{S6} - (S_{S1}\overline{S}_{S2}S_{S3}\overline{S}_{S4} \\
&+ \overline{S}_{S1}S_{S2}\overline{S}_{S3}S_{S4})S_{S5}\overline{S}_{S6})u_c - (\overline{S}_{S1}S_{S2}S_{S3}\overline{S}_{S4}S_{S5}\overline{S}_{S6})2u_c
\end{aligned}
\tag{2.28}
$$

Series connected double SM (SDSM) is proposed by Zhang et al. [126]. SDSM, specially designed for blocking dc-link fault current. It is basically two HBSMs connected using an additional switch S_5 and diode D_6, as shown in 2.6e. It can extinguish the dc arc and restore the insulation in the event of a short circuit. Furthermore, it can protect MMCs and keep them connected even under zero impedance current. Under normal operation, the switch S_5 is permanently ON and the two HBSMs are connected as independent submodules. Thus, voltage balancing, modulating and control strategies do not require any modifications [126]. The current and voltage for this mode are defined in (2.29). When the switch S_5 is OFF, the fault current passes through the diode D_6, and the anti-parallel diodes. It enables the dc-fault handling capability, as defined in (2.30).

$$
\begin{aligned}
i_{c1} =&S_{S1}\overline{S}_{S2}(\overline{S}_{S3}S_{S4} + S_{S3}\overline{S}_{S4})i_x \\
i_{c2} =& - S_{S3}\overline{S}_{S4}(S_{S1}\overline{S}_{S2} + \overline{S}_{S1}S_{S2})i_x \\
v_o =&(S_{S1}\overline{S}_{S2}\overline{S}_{S3}S_{S4} + \overline{S}_{S1}S_{S2}S_{S3}\overline{S}_{S4})u_c + (S_{S1}\overline{S}_{S2}S_{S3}\overline{S}_{S4})2u_c
\end{aligned}
\tag{2.29}
$$

$$
v_o = -(\overline{S}_{S1}\overline{S}_{S2}\overline{S}_{S3}\overline{S}_{S4}\overline{S}_{S5})u_c \tag{2.30}
$$

An asymmetrical full-bridge cross-connected double submodule (AFBCCDSM) is depicted in Fig. 2.6f [128,129]. This bi-polar and bi-direction configuration has six semiconductor switches and four capacitors. It is asymmetrical in terms of voltage levels. During the insert mode, all four capacitors can be connected in series to generate maximum voltage. However, in blocking mode, only two capacitors can be connected to provide dc fault handling capability. The currents and voltage of the AFBCCDM are defined in (2.31). The different voltages one can obtained are $-2u_c$,

0, $2u_C$ and $4u_c$.

$$
\begin{aligned}
i_{c1} =& S_{S1}\overline{S}_{S2}(S_{S3}\overline{S}_{S4} + \overline{S}_{S3}S_{S4})(S_{S5}\overline{S}_{S6} + \overline{S}_{S5}S_{S6})i_x \\
i_{c2} =& (S_{S3}\overline{S}_{S4} + \overline{S}_{S3}S_{S4})(S_{S1}\overline{S}_{S2}\overline{S}_{S5}S_{S6} - \overline{S}_{S1}S_{S2}S_{S5}\overline{S}_{S6})i_x \\
i_{c3} =& -(S_{S1}\overline{S}_{S2} + \overline{S}_{S1}S_{S2})S_{S3}\overline{S}_{S4}(S_{S5}\overline{S}_{S6} + \overline{S}_{S5}S_{S6})i_x \\
i_{c4} =& (S_{S1}\overline{S}_{S2} + \overline{S}_{S1}S_{S2})(\overline{S}_{S3}S_{S4}S_{S5}\overline{S}_{S6} - S_{S3}\overline{S}_{S4}\overline{S}_{S5}S_{S6})i_x \\
v_o =& (S_{S1}\overline{S}_{S2}(S_{S3}\overline{S}_{S4}S_{S5}\overline{S}_{S6} + \overline{S}_{S3}S_{S4}\overline{S}_{S5}S_{S6}) + \overline{S}_{S1}S_{S2}(S_{S3}\overline{S}_{S4}\overline{S}_{S5}S_{S6} \\
& - \overline{S}_{S3}S_{S4}S_{S5}\overline{S}_{S6}))2u_c + (S_{S1}\overline{S}_{S2}S_{S3}\overline{S}_{S4}\overline{S}_{S5}S_{S6})4u_c
\end{aligned}
\tag{2.31}
$$

2.4.4 Five-level submodules

Similar to the CDSM, a five-level cross-connected submodule (FLCCSM) configuration is achieved by the back-to-back cross-connection of two half-bridge submodules. As depicted in Fig. 2.7a, the additional switches S_5 and S_6 facilitate the cross-connection [95, 130]. Apart from its capabilities of producing five voltage levels, the module can provide reverse voltage similar to FBSM [130]. The currents and voltage for the same are defined in (2.32). FLCCSM has the ability to block dc fault current, as the two capacitors of SM are connected in series and can establish a blocking effect on dc-link fault current [95, 108].

$$
\begin{aligned}
i_{c1} =& (S_{S1}\overline{S}_{S2}S_{S5}\overline{S}_{S6} - \overline{S}_{S1}S_{S2}\overline{S}_{S5}S_{S6})(S_{S3}\overline{S}_{S4} + \overline{S}_{S3}S_{S4})i_x \\
i_{c2} =& (S_{S1}\overline{S}_{S2} + \overline{S}_{S1}S_{S2})(\overline{S}_{S3}S_{S4}S_{S5}\overline{S}_{S6} - S_{S3}\overline{S}_{S4}\overline{S}_{S5}S_{S6})i_x; \\
v_o =& ((S_{S1}\overline{S}_{S2}S_{S3}\overline{S}_{S4} + \overline{S}_{S1}S_{S2}\overline{S}_{S3}S_{S4})S_{S5}\overline{S}_{S6} - (\overline{S}_{S1}S_{S2} + S_{S1}\overline{S}_{S2})\overline{S}_{S3}S_{S4}\overline{S}_{S5}S_{S6})u_c \\
& + (S_{S1}\overline{S}_{S2}\overline{S}_{S3}S_{S4}S_{S5}\overline{S}_{S6} - \overline{S}_{S1}S_{S2}S_{S3}\overline{S}_{S4}\overline{S}_{S5}S_{S6})2u_c
\end{aligned}
\tag{2.32}
$$

Figs. 2.7b and 2.7c present cross-connected SM (CCSM) and parallel-connected SM (PCSM), respectively. Two FBSMs can be connected in a cross way to yield a cross-connected configuration [130]. Likewise, the FBSMs can be connected in par-

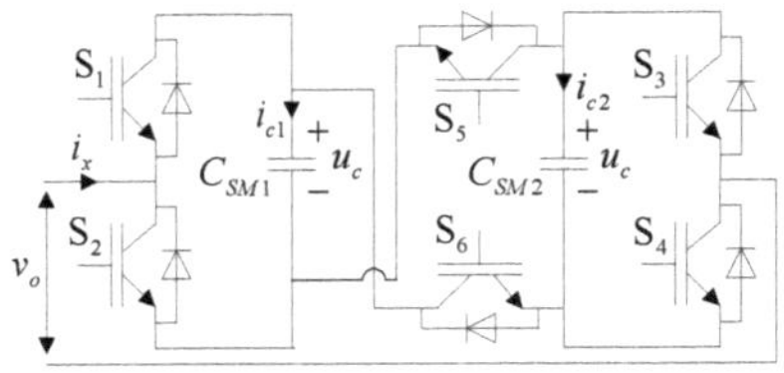

(a)

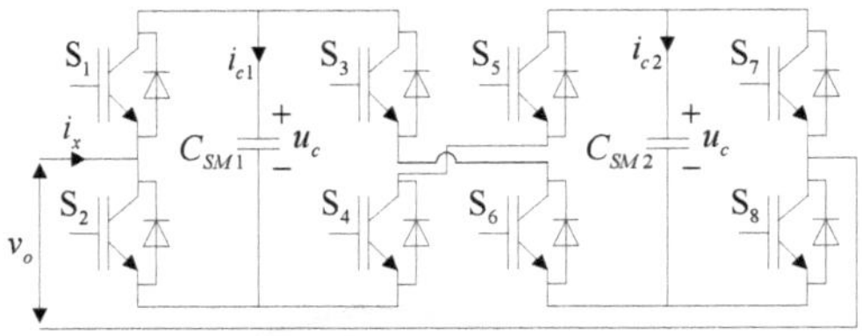

(b)

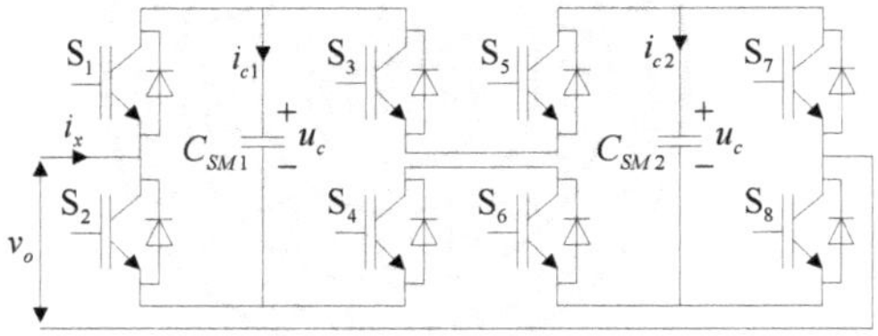

(c)

Fig. 2.7. Different five-level SM structures (a) FLCCSM, (b) cross-connected SM, (c) parallel-connected SM

allel configuration [130,131]. Both configurations have eight switches and two capacitors. Cross-connected SM is an alternative structure that can generate symmetrical bipolar output voltages and achieve high voltage levels [132]. In the parallel configuration, connecting capacitors in parallel help in the reduction of capacitor voltage ripple [123]. The currents and voltage for the cross and parallel-connected SMs are defined in (2.33) and (2.34), respectively. The presence of a full-bridge structure enables the dc fault handling capability. The maximum blocking voltage is $+2u_c$.

The semiconductor losses are twice in comparison to the FBSM. However, the voltage level also increases from three to five. Apart from these configurations, Ilves et al. [123] have proposed an alternative way of parallel connection of capacitors. The motive is to reduce the device's current rating.

$$\begin{aligned} i_{c1} =& (S_{S1}\overline{S}_{S2}\overline{S}_{S3}S_{S4}S_{S5}\overline{S}_{S6} - \overline{S}_{S1}S_{S2}S_{S3}\overline{S}_{S4}\overline{S}_{S5}S_{S6})(S_{S7}\overline{S}_{S8} + \overline{S}_{S7}S_{S8})i_x \\ i_{c2} =& (S_{S1}\overline{S}_{S2} + \overline{S}_{S1}S_{S2})(\overline{S}_{S3}S_{S4}S_{S5}\overline{S}_{S6}\overline{S}_{S7}S_{S8} - S_{S3}\overline{S}_{S4}\overline{S}_{S5}S_{S6}S_{S7}\overline{S}_{S8})i_x \\ v_o =& (S_{S1}\overline{S}_{S2}S_{S7}\overline{S}_{S8} + \overline{S}_{S1}S_{S2}\overline{S}_{S7}S_{S8})(\overline{S}_{S3}S_{S4}S_5\overline{S}_{S6} - S_{S3}\overline{S}_{S4}\overline{S}_{S5}S_6)u_c \\ &+ (S_{S1}\overline{S}_{S2}\overline{S}_{S3}S_{S4}S_5\overline{S}_{S6}\overline{S}_{S7}S_{S8} - \overline{S}_{S1}S_{S2}S_{S3}\overline{S}_{S4}\overline{S}_{S5}S_6S_{S7}\overline{S}_{S8})2u_c \end{aligned} \tag{2.33}$$

$$\begin{aligned} i_{c1} =& (S_{S1}\overline{S}_{S2}\overline{S}_{S3}S_{S4}\overline{S}_{S5}S_{S6} - \overline{S}_{S1}S_{S2}S_{S3}\overline{S}_{S4}S_{S5}\overline{S}_{S6})(S_{S7}\overline{S}_{S8} + \overline{S}_{S7}S_{S8})i_x; \\ i_{c2} =& (S_{S1}\overline{S}_{S2} + \overline{S}_{S1}S_{S2})(S_{S3}\overline{S}_{S4}S_{S5}\overline{S}_{S6}\overline{S}_{S7}S_{S8} - \overline{S}_{S3}S_{S4}\overline{S}_{S5}S_{S6}S_{S7}\overline{S}_{S8})i_x; \\ v_o =& (S_{S1}\overline{S}_{S2}\overline{S}_{S7}S_{S8} - \overline{S}_{S1}S_{S2}S_{S7}\overline{S}_{S8})(\overline{S}_{S3}S_{S4}\overline{S}_{S5}S_{S6} + S_{S3}\overline{S}_{S4}S_5\overline{S}_{S6})u_c \\ &+ S_{S3}S_4S_5S_6(S_{S1}\overline{S}_{S2}\overline{S}_{S7}S_{S8} - \overline{S}_{S1}S_{S2}S_{S7}\overline{S}_{S8})2u_c \end{aligned} \tag{2.34}$$

2.4.5 Special submodules

Riar et al. [133] suggest the modification in SMs for balancing the voltage using the concept of inductive power transfer. It is termed as a "cell with resonant inverter for inductive power transfer (CRI-IPT)". The cell configuration is shown in Fig. 2.8a [94, 133]. In this topology, the inductor and capacitor form a resonant circuit. It can be seen as a modification of the full-bridge submodule.

Perez et al. [134] propose the idea to replace the voltage source dc-link-based cell topology with a current source dc-link-based cell topology. The concept is presented in Fig. 2.8a. The current source SM is a half-bridge with a capacitor at ac terminal and an inductor at the dc terminal. Switches must be operated in a way that maintains an interruption-free dc current flow. In the comparison of traditional voltage source SMs, the current source SMs can achieve higher voltage and power ratings [94, 134].

Fig. 2.9 depicts three special submodules for the hybrid structure of the MMC. These SMs are designed to be used at the middle point of the MMC leg, where the upper and lower legs meet. When connecting with the system, the upper terminal of the middle SM gets connected to the upper leg and the lower terminal gets connected to the lower arm. The third terminal gets connected to the ac network [135].

Among all submodule configurations, HBSM is considered to be the most suited. Its simple architecture of two switches results in lower losses and higher efficiency than other submodules [7, 95, 136]. Table 2.1 shows the enhancement/improvement in submodule characteristics in terms of the number of semiconductor components and capacitors, losses and efficiency. Apart from that, the mechanical design of the submodule, internal fault protection system, capacitors requirement and control issues under high power applications are also major concerns. Thus, a trade-off between reliability/functionalities and complexity poses a serious challenge to the approaches to finding the optimal power submodules [9, 113, 125].

2.5 Key Issues and Challenges of MMC

Besides all the advantages, MMC possess several issues and control challenges that provide an opportunity to researchers for further improvement in existing architecture and control schemes. These challenges include design constraints, control issues,

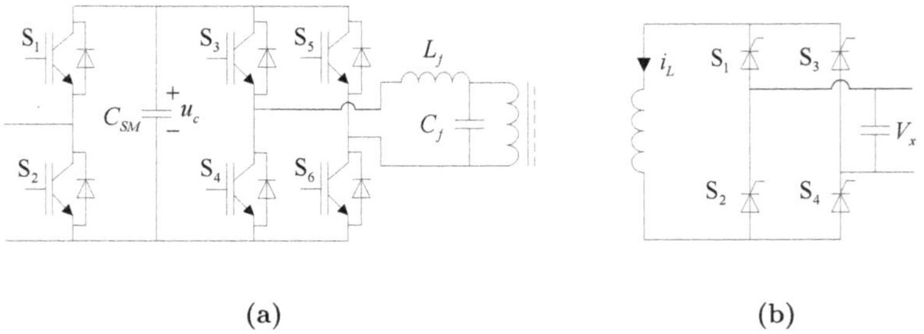

(a) (b)

Fig. 2.8. Special SM structures (a) cell with resonant inverter for inductive power transfer, (b) current source SM

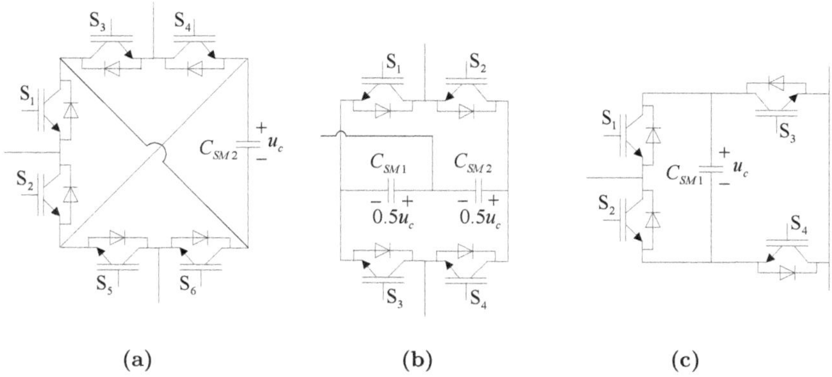

Fig. 2.9. Middle SM structures (a) type 1, (b) type 2, (c) type 3

pre-charging of capacitors and start-up process, modelling, etc. [25, 94, 95, 113, 114].

2.5.1 Design constraints of MMC

In the MMC design, the primary design constraints are:

- Size of the inductor
- Size of capacitor
- Semiconductor losses
- Reliability

MMC's inductor filters high-frequency harmonics in arm current and serves as a limiter for dc-side fault current. As a result, short circuit dc current and arm ripple current have a significant impact on its dimension. The capacitor's size is determined by voltage ripple and cost. The MMC process generates ripples of the fundamental, second, and all higher-order frequency components. An analysis of these ripples is required prior to selecting the capacitor. A typical MMC configuration has over 200 SMs per arm. These SMs suffer from both switching and conduction losses. As a result, submodules with fewer switches are desirable for minimising losses. However, the demand for dc fault handling capability and the desire for reduced capacitor size results in a trade-off. The issue of reliability is directly related to the amount of redundancy. An MMC with more redundant SMs is more dependable because it can keep the power on even if a few SMs fail. Taking these constraints into account, numerous hybrid architectures of MMCs and SMs are developed. However, most of them give up one or more of MMC's advantages, such as modularity, efficiency, dc side fault ride-through capabilities, and so on.

2.5.2 Control challenges of MMC

The MMC operation is a multi-objective function. Its control can be sub-divided into multiple parallel processes. The primary processes can be listed as follows:

- Power flow control
- Pole to pole dc voltage control
- Minimising the circulating current
- SM capacitors voltage balancing
- Minimizing voltage ripples
- Active and reactive power decoupling
- The dc-fault handling

Despite the fact that numerous control methods have been developed to address the aforementioned concerns, the need for simultaneous control of all parameters without sacrificing efficiency remains a concern. The search for the most optimal control strategy is still going on.

2.5.3 Capacitor pre-charging and start-up of MMC

MMC capacitors must be pre-charged for a quick and smooth start-up. Several alternatives have been proposed, but they all require either expensive dc circuit breakers or complex control techniques, or both.

2.5.4 Modelling of MMC

MMC's practical structure has over 200 SMs per arm. Furthermore, each SM contains a number of semiconductor devices and capacitors. In such a case, system analysis becomes challenging because taking into account all of the details would necessitate a large amount of processing and storage capacity. On the other side, speeding up the procedure would mean sacrificing the quantity of detail. As a result, while modelling an MMC, the trade-off between the two is inevitable.

2.6 Real-Time Digital Simulator (RTDS)

The real-time digital simulator (RTDS) has been developed by RTDS technologies [137]. RTDS Technologies developed the RTDS Simulator with highly customised proprietary hardware and software for real-time power system simulation. This study makes use of the RTDS facility at the advanced power system research lab of the Indian institute of technology Kanpur [138]. As shown in Fig. 2.10, the RTDS in this facility consists of six "racks." Each rack contains a number of parallel processing cards, including simulation running cards, inter-rack communications (IRC) cards, workstation interface (WIF) cards, and input-output (I/O) cards. It employs the PB5 processor card, the most recent generation of processor card created for the RTDS Simulator. Each PB5 is equipped with two PowerPC reduced instruction set computer (RISC) processors operating at 1.7 GHz.

The two components of RTDS software are the RTDS operating system (OS) and compiler. The compiler runs on RTDS racks, whereas the OS is distributed. A portion of the OS runs on the host computer of the RTDS rack, while the remaining portion runs on the giga transceiver workstation interface (GTWIF) card. The RTDS compiler converts simulations to executable machine code. The compiler also handles parallel processing among digital signal processors, allocation of memory, and communication. The graphic user interface-based software tool of RTDS is RSCAD. It provides RTDS with a comprehensive graphical interface. The draft

Fig. 2.10. RTDS facility at the advanced power system research lab of Indian institute of technology Kanpur.

module of RSCAD is primarily utilised for circuit assembly and parameter entry.

2.7 Conclusion

For voltage source converters, modular multi-level converter (MMC) technology is rapidly evolving. It has only been practically implemented in high voltage dc transmission. However, there is a significant amount of effort being done around the world to apply it in a variety of other sectors. The technology is evolving not only in terms of converter circuits but also in terms of its fundamental building block, i.e. submodule. Because of its simple and efficient structure, the HBSM is the most commonly used submodule; however, it lacks the capacity to prevent dc-link short circuit fault currents. FBSM is capable of limiting the flow of dc-link fault current, but its larger number of components results in higher losses. So, the process of finding optimal submodule structure has resulted in numerous SM topologies, and study on the subject is ongoing.

Chapter 3

Development of EMT Model and Stability Analysis of MMC

3.1 Introduction

Modular multilevel converter (MMC) technology has evolved significantly during the last decade. The modular structure of the MMC provides it with an edge over alternative converter topologies. Many studies have made substantial contributions to understanding the MMC's modelling and operation. The literature contains the comparative study of the various types of electromagnetic transient (EMT) models [58, 59]. Likewise, a few studies include the MMC's steady-state analysis [65, 66, 68]. The interaction of current and voltage over the sub-module capacitor (SM capacitor) has been explored. Because of the MMC's intricate structure, studies focus on specific aspects while neglecting others. The authors of [60] presented a method for developing a detailed small-signal model as well as a stability analysis. Their strategy improves stability but results in additional power loss.

In this chapter, an EMT model of the MMC is developed using the real-time digital simulator. A detailed small-signal model has also been developed in MATLAB. The EMT model has been validated in comparison to the small-signal model. The stability study has been performed to assess the range of controller variables. An investigation of the necessity and impact of the circulating current controller (also known as the auxiliary controller) on the dynamics of the MMC is also presented. The EMT model developed in this chapter has been used in the subsequent chapters.

3.2 MMC Operation

The MMC architecture is described in Chapter 2. MMC maintains the fundamental frequency voltages $v_{tj(j=abc)}$ at the ac terminal and V_{dc} at the dc terminal of the MMC leg as shown in Fig. 2.1. The ac and dc terminal voltages can be seen as multilevel voltage synthesis. As the number of SMs increases, the waveform becomes finer and finer. Furthermore, the terminal voltage v_t is a function of the upper and lower arm voltages (v_u and v_l). The upper and lower arm voltages are synthesised by varying the number of SMs inserted. This can be accomplished with the use of an appropriate switching function. The arm voltage relationship to the number of SMs (N), the SM capacitor voltage u_c, and the switching function S is defined by (3.1). The upper and lower arms are denoted by the subscripts u and l, respectively. Since, with the exception of the phase difference, all three phases act identically, the phase information has been omitted for simplicity. Furthermore, in order to synthesise the constant dc voltage V_{dc}, the total number of SMs placed in each phase leg must be constant, as defined in (3.2).

$$v_u = N * S_u * u_{cu} \quad \text{and} \quad v_l = N * S_l * u_{cl} \tag{3.1}$$

$$N * S_u + N * S_l = N \tag{3.2}$$

$$\begin{aligned} i_u &= \frac{1}{3} I_{dc} - \frac{1}{2} I_s \sin(wt + \alpha_s) + I_{cir} \sin(2wt + \alpha_c) \\ i_l &= \frac{1}{3} I_{dc} + \frac{1}{2} I_s \sin(wt + \alpha_s) + I_{cir} \sin(2wt + \alpha_c) \end{aligned} \tag{3.3}$$

The upper and lower arm currents, i_u and i_l, are a combination of the dc (I_{dc}), fundamental ac (i_s), and an even-order circulating current (i_{cir}), as defined in (3.3). The fourth and higher-order components are omitted due to their negligible magnitudes. The I_s and α_s represent the amplitude and phase angle of the fundamental frequency current, respectively. Likewise, I_{cir} and α_c are the amplitude and phase angle of the second-order circulating current, respectively.

The fundamental ac component of the arm current links the MMC to the ac network. Similarly, the dc component connects it to the dc network. The second-order component, on the other hand, flows between the phase legs, resulting in power losses. The essential dc and fundamental components can be maintained while suppressing the unwanted second-order component using a suitable arm switching function as defined in (3.4). S_u and S_l are average arm switching functions for the upper and lower arms, respectively. Similar to arm current, it also is composed of three components. The dc and fundamental components establish the respective currents in the arm. The second-order component suppresses the circulating current. M is the fundamental component's modulation index, which is generated from the primary controller (also known as the fundamental current controller or output current controller), as described in the following sections. Similarly, the V_{cir} is the auxiliary controller's (also known as circulating current controller) output. The phase angles of the fundamental and second-order components are represented by β_f and β_c, respectively.

$$\begin{aligned} S_u &= \frac{\frac{1}{2}V_{dc} - \frac{1}{2}MV_{dc}\sin(wt+\beta_f) + V_{cir}\sin(2wt+\beta_c)}{V_{dc}} \\ S_l &= \frac{\frac{1}{2}V_{dc} + \frac{1}{2}MV_{dc}\sin(wt+\beta_f) + V_{cir}\sin(2wt+\beta_c)}{V_{dc}} \end{aligned} \tag{3.4}$$

$$S_u.i_u = C.\frac{du_{cu}}{dt} \quad \text{and} \quad S_l.i_l = C.\frac{du_{cl}}{dt} \tag{3.5}$$

The SM capacitor charges or discharges in accordance with the arm current flowing through it, which, in turn, has an effect on the SM capacitor voltage. The current flow is determined by the mode of SM. Furthermore, the switching function controls the SM's mode. This relationship between the arm current, SM capacitance, SM capacitor voltage, and average switching function is defined by (3.5). Moreover, as defined by (3.6), the SM capacitor voltage is the sum of the dc component $\overline{u}_c$ and the ac ripple of fundamental and harmonic frequency components. The magnitude and angle of the kth order ripple are denoted by u_{ck} and θ_k, respectively. Fourth- and higher-order harmonic ripples can be ignored [65].

$$u_c = \overline{u}_c + \sum_{k=1}^{\infty} u_{ck} \sin(kwt + \theta_k) \tag{3.6}$$

3.3 Modelling of MMC

In this section, the electromagnetic transient (EMT) model has been developed in the real-time digital simulator (RTDS), and the small-signal model has been developed in MATLAB.

3.3.1 EMT model of MMC for HVDC application

Fig. 3.1 illustrates the conceptual block diagram for the integration of MMC, with the ac and dc side networks and controllers, for which an EMT model is constructed in RTDS. MMC is connected to the ac network by transformer inductance L_t at the point of common coupling (PCC). The dc terminal is coupled to a dc network. The controllers which are implemented in the d-q coordinates, receive measurements of the PCC voltage, phase currents, arm currents, and SM capacitor voltage (shown by dashed lines). When switch S1 is connected to terminal P1, the MMC operates in P-Q mode, and when the switch is connected to terminal P2, the MMC operates in V-Q mode. In the P-Q mode, the objective is to control the active power exchange between the ac and dc network and provide reactive power support to the ac network. In the V-Q mode of operation, the objective of MMC is to maintain the reference voltage at the dc terminal and provide reactive power support to the ac network. The reactive power support is required to maintain the desired ac voltage at the PCC.

The control philosophy has been implemented in d-q coordinates using the dy-

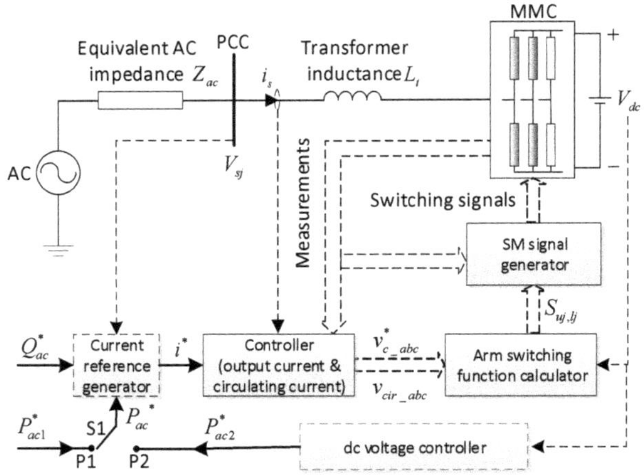

Fig. 3.1. Integration of MMC with ac grid, dc network and control

namic phasors. The current reference generator block creates references for the current controller based on active and reactive power references and PCC voltage measurement. As shown in Fig. 3.2, the controller block is a combination of two controllers that operate in parallel, namely an output current (or fundamental current) controller and a circulating current (or auxiliary) controller.

The output current controller generates voltage references using the current reference and the measured value of phase current. The phase current is measured directly by deploying a transducer in each phase on either side of the transformer. They can also be mathematically deduced from the arm current measurements using (3.7). The circulating current controller generates voltage references to cancel the circulating current based on the measured value from the MMC block. The measurements are obtained using (3.7). The reference signal for the circulating current controller is zero. The arm switching function calculator then uses the outputs of both controllers to calculate the average switching functions for each arm. The switching function and voltages of each SM are used to generate the switching signal for each SM via the SM signal generator. Additionally, this block is in charge of voltage balancing across all SMs. The MMC is operated in P-Q mode by directly feeding the P_{ac}^* and Q_{ac}^*, whereas in V-Q mode, the P_{ac}^* is derived using a dc voltage controller. To achieve accurate simulation in the developed RTDS model, the converter architecture is designed within a small time step bridge box (rtds_vsc_BRIDGE_BOX). The arm structures of the MMC are realised using the "VSC Model Type: MMC5". It gives the total arm current of each arm. It can be used to deduce various current components mathematically as defined in (3.7).

$$\begin{aligned} i_s &= I_s \sin(wt + \alpha_s) = i_l - i_u \\ i_{cir} &= I_{cir} \sin(2wt + \alpha_c) = \frac{1}{2}(i_l + i_u) - \frac{1}{3} I_{dc} \\ I_{dc} &= \frac{1}{2} \sum_{j=a,b,b} (i_l + i_u) \end{aligned} \tag{3.7}$$

In terms of the voltage, the available data includes the average SM capacitor voltage and the arm voltage (including as well as excluding the arm resistance and inductance). Furthermore, the voltage of the blocked SM string, the blocked section of the arm, and the status of the diode are also accessible. The "VSC Model Type: MMC5" block includes the SM capacitor voltage balancing approach. The ac and dc sides controller are designed outside the small time step bridge box.

3.3.2 State-space model of MMC

The MMC's state-space modelling has been implemented in MATLAB. For this, the concept of dynamic-phasor is employed. Three-phase alternating current components that vary in time are converted into the equivalent d-q component. The transformation matrix for the kth order component is given in (3.8). For the fundamental and second-order components, the values of k are 1 and 2, respectively. Using (3.8), with $k = 1$, the phase voltage $v_{cj(j=a,b,c)}$ in a-b-c coordinates can be defined in d-q coordinates as:

$$P_k = \frac{2}{3}\begin{bmatrix} \cos(kwt) & \cos(kwt - \frac{2k\pi}{3}) & \cos(kwt + \frac{2k\pi}{3}) \\ \sin(kwt) & \sin(kwt - \frac{2k\pi}{3}) & \sin(kwt + \frac{2k\pi}{3}) \end{bmatrix} \tag{3.8}$$

$$V_{cd} = \frac{1}{2}V_{dc}M\sin\beta_f \text{ and } V_{cq} = \frac{1}{2}V_{dc}M\cos\beta_f \tag{3.9}$$

In terms of the voltage, there exists symmetry. Each of the upper and lower arms (as well as the SM capacitor) has the identical dc voltage component and even harmonic voltage ripples, with the exception of an odd harmonic ripple with the opposite sign in both arms. The voltages in all three phases are separated by 120 degrees. As a result, by observing the dynamics of arm voltage and SM capacitor voltages in one arm, one can infer the dynamics of all six arms and their SM capacitors in the other arms.

Substituting (3.3) and (3.4) in (3.5), the dynamics of the SM capacitor voltage, which is made up of both dc and ac ripples can be obtained. By segregating the components of each harmonic and transforming them using (3.8), the dynamic of SM in d-q coordinates can be derived. Thus, we obtain $\frac{d}{dt}\overline{u}_c$, $\frac{d}{dt}u_{c1d}$, $\frac{d}{dt}u_{c1q}$, $\frac{d}{dt}u_{c2d}$, $\frac{d}{dt}u_{c2q}$, $\frac{d}{dt}u_{c3d}$, and $\frac{d}{dt}u_{c3q}$, the dynamics of quasi-static dc component, 1st, 2nd, and 3rd order voltage ripple components in d-q coordinate, ignoring 4th and higher order. Applying the KVL for different frequency components gives the dynamics of the dc, fundamental and second order circulating current, respectively as given in (3.10), (3.11) and (3.12).

$$\frac{d}{dt}I_{dc} = \frac{3}{2L_{arm}}(V_{dc} - 2v_{u_{dc}}) - \frac{R_{arm}}{L_{arm}}I_{dc} \tag{3.10}$$

$$\frac{d}{dt}i_s = \frac{1}{L_{eq}}(v_s + v_{u1}) - \frac{R_{eq}}{L_{eq}}i_s \tag{3.11}$$

$$\frac{d}{dt}i_{cir} = -\frac{1}{L_{arm}}v_{u2} - \frac{R_{arm}}{L_{arm}}i_{cir} \tag{3.12}$$

where, $L_{eq} = L_t + L_{arm}$, v_{udc} is upper arm dc voltage; v_{u1} is fundamental and v_{u2} is 2nd order arm voltage ripple component, and can be extracted from arm voltage equation, obtained by substituting (3.4) and (3.6) in (3.1). Further, using the transformation matrix (3.8), alternating currents of (3.11), and (3.12), yields

$\frac{d}{dt}I_{sd}$, $\frac{d}{dt}I_{sq}$, $\frac{d}{dt}I_{cird}$, and $\frac{d}{dt}I_{cirq}$, respectively d-q component dynamics of fundamental output and second order circulating current. Authors in [60] have developed a detailed state-space model of MMC. Small-signal model of MMC is obtained by linearising the 12th order MMC model as follows:

$$\mathbf{\Delta\dot{x}} = \mathbf{A_{MMC}\Delta x} + \mathbf{B_{MMC}\Delta u} \tag{3.13}$$

where,

$$\begin{aligned}\mathbf{\Delta x} =& [\mathbf{\Delta\overline{u}_c}, \mathbf{\Delta u_{c1d}}, \mathbf{\Delta u_{c1q}}, \mathbf{\Delta u_{c2d}}, \mathbf{\Delta u_{c2q}}, \mathbf{\Delta u_{c3d}}, \mathbf{\Delta u_{c3q}}, \mathbf{\Delta I_{dc}}, \mathbf{\Delta I_{sd}}, \mathbf{\Delta I_{sq}},\\ & \mathbf{\Delta I_{cird}}, \mathbf{\Delta I_{cirq}}]^T\\ \mathbf{\Delta u} =& [\mathbf{\Delta V_{cird}}, \mathbf{\Delta V_{cirq}}, \mathbf{\Delta V_{cd}}, \mathbf{\Delta V_{cq}}, \mathbf{\Delta V_{sd}}, \mathbf{\Delta V_{sq}}, \mathbf{\Delta V_{dc}}, \mathbf{\Delta}\omega]^T\end{aligned}$$

where $\mathbf{A_{MMC}}$ and $\mathbf{B_{MMC}}$ are the state and input matrix; the $\mathbf{\Delta x}$, and $\mathbf{\Delta u}$ denote the state and input vectors. Various variables of the input vectors serve as an interface between the MMC and the rest of the system's components. The voltages $\mathbf{\Delta V_{cd}}$ and $\mathbf{\Delta V_{cq}}$ serve as interfaces with the primary controller, while the voltages $\mathbf{\Delta V_{cird}}$ and $\mathbf{\Delta Vcirq}$ serve as interfaces with the circulating current controller. The voltages $\mathbf{\Delta V_{sd}}$ and $\mathbf{\Delta V_{sq}}$ connect the MMC to the alternating current system, while $\mathbf{\Delta V_{dc}}$ links it to the direct current system, and $\mathbf{\Delta}\omega$ facilitates the interface with the phase-locked loop (PLL).

3.3.3 Controllers for MMC operation

Fig. 3.2 illustrates the block diagram of the controllers. As discussed earlier, the input reference for the output current controller is an active and reactive power reference. The reference for the circulating current controller is zero, as the objective is to suppress the unwanted circulating current. The blocks $\frac{1}{1+sT_i}$ and $\frac{1}{1+sT_v}$ represent the current and voltage transducers, respectively. T_i and T_v are respective time constants. Both PI-based controllers operate in parallel. The controllers are realised using the proportional-integral (PI) control philosophy. From the PI blocks of the control structure, the following relations can be derived:

$$\left.\begin{aligned} y_{1d} &= \int (I_{sd}^* - I_{sdm})dt \\ y_{1q} &= \int (I_{sq}^* - I_{sqm})dt \\ y_{2d} &= \int (I_{cird}^* - I_{cirdm})dt \\ y_{2q} &= \int (I_{cirq}^* - I_{cirqm})dt \end{aligned}\right\} \rightarrow \begin{cases} \dfrac{dy_{1d}}{dt} = I_{sd}^* - I_{sdm} \\ \dfrac{dy_{1q}}{dt} = I_{sq}^* - I_{sqm} \\ \dfrac{dy_{2d}}{dt} = I_{cird}^* - I_{cirdm} \\ \dfrac{dy_{2q}}{dt} = I_{cirq}^* - I_{cirqm} \end{cases} \tag{3.14}$$

I_{sd}^*, I_{sq}^*, I_{cird}^*, are I_{cirq}^* are the reference value of d-q current components, corresponding to output and circulating currents, respectively. I_{sd}, I_{sq}, I_{cird}, and I_{cirq} are actual values of respective components. These are measured through the transducer as I_{sdm}, I_{sqm}, I_{cirdm}, and I_{cirqm} and fed to the controller as a feedback signal. Similarly, V_{sd} and V_{sq} are d-q components of PCC voltage, and V_{sdm} and V_{sqm} are corresponding measured values. The relation between actual and measured values can be defined as in (3.15).

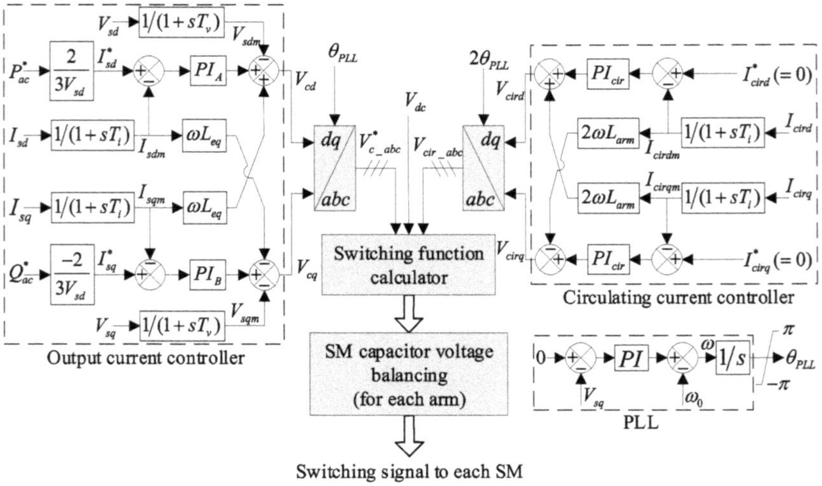

Fig. 3.2. Block diagram of controllers

$$\begin{aligned}
\frac{d}{dt}I_{sdm} &= \frac{1}{T_i}(I_{sd} - I_{sdm}) \\
\frac{d}{dt}I_{sqm} &= \frac{1}{T_i}(I_{sq} - I_{sqm}) \\
\frac{d}{dt}V_{sdm} &= \frac{1}{T_v}(V_{sd} - V_{sdm}) \\
\frac{d}{dt}V_{sqm} &= \frac{1}{T_v}(V_{sq} - V_{sqm}) \\
\frac{d}{dt}I_{cirdm} &= \frac{1}{T_i}(I_{cird} - I_{cirdm}) \\
\frac{d}{dt}I_{cirqm} &= \frac{1}{T_i}(I_{cirq} - I_{cirqm})
\end{aligned} \tag{3.15}$$

From the phase-locked loop (PLL), the following relations have been derived:

$$\frac{d}{dt}y_p = 0 - V_{sq} \tag{3.16}$$

$$\frac{d}{dt}\theta_{PLL} = \omega_0 + K_{p_{PLL}}V_{sq} - K_{i_{PLL}}y_p \tag{3.17}$$

where, $K_{p_{PLL}}$ and $K_{i_{PLL}}$ are proportional and integral constants of PLL's PI block. Current reference I_{sd}^* and I_{sq}^* are derived from (3.18), by maintaining $V_{sq} = 0$, via PLL. P_{ac}^* and Q_{ac}^* are, respectively, active and reactive power references. Further, the PCC voltage is defined as in (3.19).

$$\begin{aligned}
P_{ac}^* &= \frac{3}{2}(V_{sd}I_{sd}^* + V_{sq}I_{sq}^*) \\
Q_{ac}^* &= \frac{3}{2}(V_{sq}I_{sd}^* - V_{sd}I_{sq}^*)
\end{aligned} \tag{3.18}$$

$$\left.\begin{aligned}
v_{sa}(t) &= V_{sm}\sin(\omega t + \alpha) \\
v_{sa}(t) &= V_{sm}\sin(\omega t + \alpha - \tfrac{2\pi}{3}) \\
v_{sa}(t) &= V_{sm}\sin(\omega t + \alpha + \tfrac{2\pi}{3})
\end{aligned}\right\} \rightarrow \left\{\begin{aligned}
V_{sd} &= V_{sm}\sin(\alpha) \\
V_{sq} &= V_{sm}\cos(\alpha)
\end{aligned}\right. \tag{3.19}$$

where, v_{sa}, v_{sb}, and v_{sc} are the instantaneous fundamental frequency voltages of phases a, b and c, respectively, at PCC. V_{sm} and α are the magnitude and phase angle of voltages.

Combining the controller, PLL, and ac system models with the MMC model via interface variables enables the derivation of a state-space model for the entire system, as illustrated in Fig. 3.1. The small-signal model is obtained by linearising the same. The entire system's small-signal model with a circulating current controller is found to be of 24th order, as defined in (3.20).

$$\mathbf{\Delta \dot{X}} = \mathbf{A\Delta X} + \mathbf{B\Delta U} \tag{3.20}$$

$$\begin{aligned}\mathbf{\Delta X} = [&\mathbf{\Delta \overline{u}_c}, \mathbf{\Delta u_{c1d}}, \mathbf{\Delta u_{c1q}}, \mathbf{\Delta u_{c2d}}, \mathbf{\Delta u_{c2q}}, \mathbf{\Delta u_{c3d}}, \mathbf{\Delta u_{c3q}}, \mathbf{\Delta I_{dc}}, \mathbf{\Delta I_{sd}}, \mathbf{\Delta I_{sq}},\\ &\mathbf{\Delta I_{cird}}, \mathbf{\Delta I_{cirq}}, \mathbf{\Delta I_{sdm}}, \mathbf{\Delta I_{sqm}}, \mathbf{\Delta V_{sdm}}, \mathbf{\Delta V_{sqm}}, \mathbf{\Delta I_{cirdm}}, \mathbf{\Delta I_{cirqm}},\\ &\mathbf{\Delta y_{1d}}, \mathbf{\Delta y_{1q}}, \mathbf{\Delta y_{2d}}, \mathbf{\Delta y_{2q}}, \mathbf{\Delta y_{p}}, \mathbf{\Delta \theta_{PLL}}]^T\\ \mathbf{\Delta U} = [&\mathbf{\Delta P_{ac}}^*, \mathbf{\Delta Q_{ac}}^*, \mathbf{\Delta V_{dc}}, \mathbf{\Delta \omega}]^T\end{aligned}$$

Without the circulating current controller, the system order reduces to 20, as the variables $\mathbf{\Delta I_{cirdm}}$, $\mathbf{\Delta I_{cirqm}}$, $\mathbf{\Delta y_{2d}}$, and $\mathbf{\Delta y_{2q}}$ are eliminated from state vector $\mathbf{\Delta X}$.

3.4 Model Validation

Both the EMT and the small-signal models, built using RTDS and MATLAB, respectively, are validated against each other. The two scenarios considered for the validation are with and without a circulating current controller. Fig. 3.3 depicts the dynamic response of both models without the circulating current controller. The response for the case with the circulating current controller is presented in Fig. 3.4. Considering the two scenarios also helps to analyse the impact of the circulating current controller. The EMT model is represented by the solid (blue) line. The dashed (red) line represents the small-signal model. The PI parameters in Table 3.1 are used to implement small-signal and EMT models. The EMT (RTDS) model agrees with the small-signal (MATLAB) model.

In both scenarios, initially, the system is in steady-state operation. It is transferring 400 MW of active power from the ac to the dc network, while the reactive

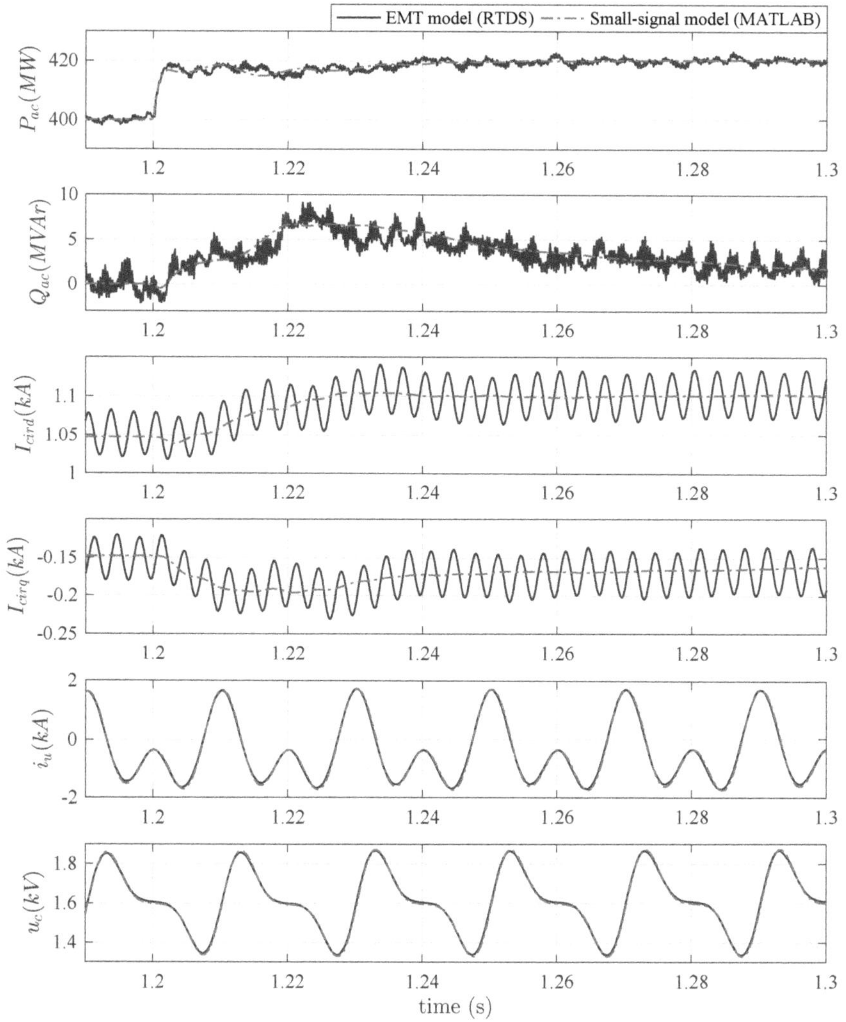

Fig. 3.3. Validation of small-signal and EMT models without circulating current controller

support is 0 MVAr. Active power is increased to 420 MW at time instant $t = 1.2$ seconds. As presented in Fig. 3.3, both models' active power (P_{ac}) plots overlap. The EMT plots have high-frequency components. The small-signal plots correspond to the EMT plot's mean value. As active and reactive powers are coupled, a quick variation in the reactive power (Q_{ac}), is also evident. The plots of the EMT model overlap with the plot of the small-signal model. The maximum value of Q_{ac} for the EMT model is 9 MVAr. The same is 6.7 MVAr for the small-signal model.

For I_{cird}, the EMT model gives an oscillating plot around a mean value, which is the same as the plot of the small-signal model. Initially, this value is 1.05 kA. It rises in response to the change in active power and settles at 1.1 kA. A similar observation can be made on the I_{cirq} plot too. The mean value of EMT and the small-signal model's value are initially set to -0.15 kA. It rises to -0.197 kA following the transient and settles at -0.16 kA. i_u represents the plots for upper arm current. The plots for both models overlap perfectly. As mentioned in previous sections, the upper arm current includes dc, fundamental, and second-order circulating currents. The effect of circulating current is evident, as it causes the distortion in the arm current waveform. Likewise, the plots of the SM capacitor voltage for both models overlap accurately. The dc shift is due to the dc component, while the distorted waveform primarily includes fundamental and second-harmonic components. From these plots, it can be concluded that the developed models validate each other for the case without circulating current control.

Table 3.1: PI parameters of the controller

Parameters	Values
K_{pA} and K_{pB}	25
K_{iA} and K_{iB}	500
K_{pcir}	30
K_{icir}	1000

Fig. 3.4, compares the plot of EMT and the small-signal model of the MMC system, including the circulating current control. Similar to the previous scenario, plots of both model overlaps accurately. Moreover, the impact of the circulating current control is also visible. On the reactive power plot, the maximum values attained by the EMT and the small-signal plot are 6 and 4.1 MVAr, respectively. Compared to the plot of Fig. 3.3, these values are significantly small. In this case, also, the plots of the I_{cird} and I_{cirq} from both models validate each each-other. Since, the circulating current controller suppresses the undesired second-order

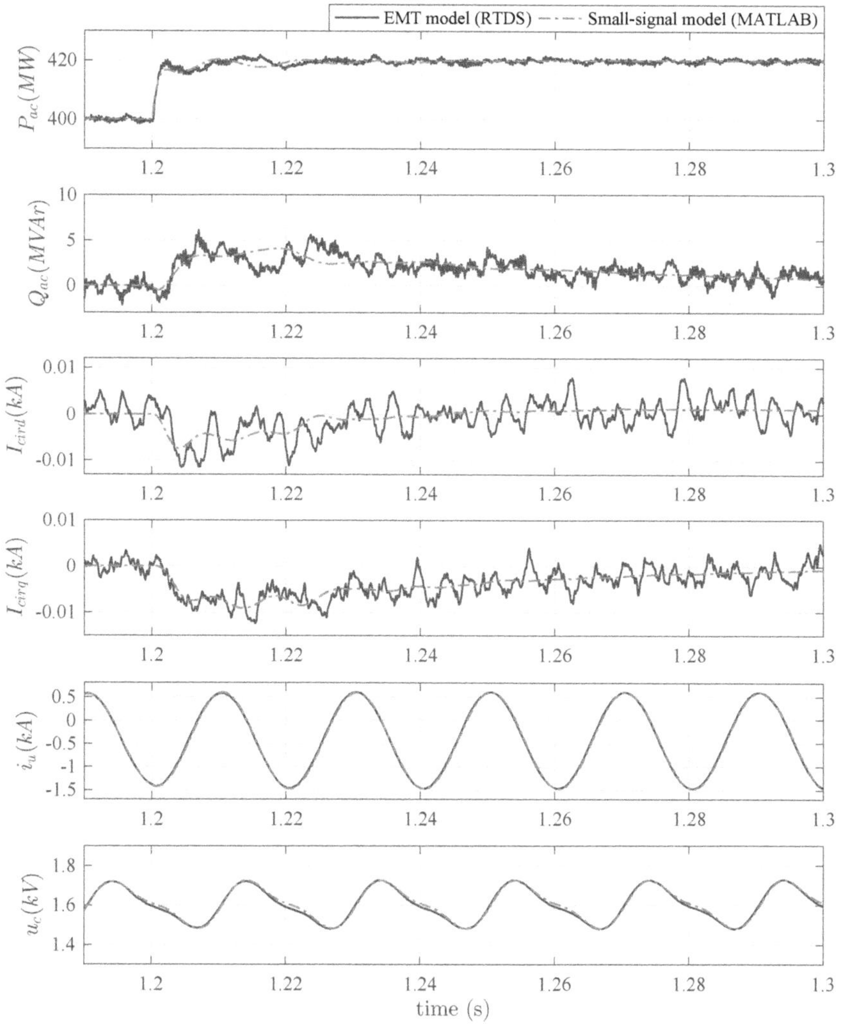

Fig. 3.4. Validation of small-signal and EMT models with the circulating current controller

current, the initial values of both I_{cird} and I_{cirq} are zero. Following the transient, they rise momentarily up to -0.01 kA and settle back to zero. With suppression of the circulating current, the arm current consists of dc and fundamental components only, as shown by the plots of the upper arm current. Both EMT and small-signal model generates identical arm current. The same can be said about the SM capacitor voltage as well. In the presence of the circulating current controller, the waveforms are free from second-order distortion.

3.5 Stability Analysis

Stability is a critical component of every system. The eigenvalue and root locus of the MMC small-signal model must be obtained to investigate the same. The MMC is designed for 500 MVA, 166 kV line to line ac voltage and ±160 kV pole to pole dc voltage. To assert stability, the operating parameters are set to 500 MW and 0 MVAr (i.e., full load). The voltages are kept at the rated values. Table 3.2 presents the system's eigenvalues at the set-point.

Table 3.2: Eigenvalues for stability analysis

	Case A			Case B			
Modes	Eigenvalues	f(Hz)	ζ	Eigenvalues	f(Hz)	ζ	Participating variables
1	-3670±480	76.3	0.99	-3669±480	76.3	0.99	I_{sdm}, I_{sqm}, I_{sd}, I_{sq}
2	-1220±199	31.6	0.99	-1235±183	29.1	0.99	I_{sq}, I_{sd}, I_{sqm}, I_{sdm}
3	-147.2±737	117	0.19	-159.5±764	122	0.20	u_{c2d}, u_{c2q}
4	-844.2±136	21.6	0.99	-798.8±204	32.5	0.97	I_{cirq}, I_{cird}
5	-20.15±948	151	0.02	-38.29±948	151	0.04	u_{c3x}, u_{c3y}
6	-82.85±551	87.7	0.15	-85.05±544	86.5	0.11	I_{dc}, u_{c1d}, $\overline{u}_c$, u_{c1q}
7	-66.13±264	42.0	0.24	-56.10±270	42.9	0.20	u_{c1q}, $\overline{u}_c$, u_{c1d}, I_{dc}
8	-18.57±3.44	0.56	0.98	-21.92±4.75	0.76	0.98	y_{1d}, y_{1q}, y_{2d}, y_{2q}
9	-1355374	0	1	-1355374	0	1	θ_{PLL}
10	-10	0	1	-10	0	1	y_p
11	-5000	0	1	-5000	0	1	V_{sdm}
12	-5000	0	1	-5000	0	1	V_{sqm}
13				-5008±0.01	0.001	1	I_{cirdm}, I_{cirqm}
14				-23.79±9.97	1.59	0.92	y_{2d}, y_{2q}, y_{1d}, y_{1q}

The two situations under consideration for study are without and with a circulating current controller (Case A and Case B). Cases A and B have a total of 12 and 14 operating modes, respectively. Each mode is described in detail, including the oscillation frequency (Hz), damping ratio (ζ), and primary participating variables. The modes associated with the ac current (both fundamental and circulating) and current transducers (i.e. 1st, 2nd and 4th) have the damping ratio of almost unity.

The eigenvalues, although fluctuating in nature, are far from the imaginary axis. It should be noticed that all of the least damped modes are associated with the capacitor voltage ripple (i.e. 3rd, 5th, 6th, and 7th). Because of the incorporation of a circulating current controller, damping for two of the least damped modes (i.e. 5th and 3rd) can be improved, but dampings for modes 6th and 7th are marginally degraded. Furthermore, the primary variables associated in these modes are dc and fundamental components of the SM capacitor voltage and dc current. The auxiliary controller causes the eigenvalues of the 2nd, 5th, and 6th modes to shift away from the imaginary axis. The eigenvalue of the 7th mode, on the other hand, shifts marginally towards the imaginary axis. The eigenvalues of the 5th, 8th, and 10th can be considered the closest to the imaginary axis. These are primarily regulated by the third-order SM capacitor voltage, controllers (both primary and auxiliary), and the PLL, respectively. While the damping ratio of the 8th mode is close to unity, the damping ratio of the 10th mode is unity. With the addition of the auxiliary controller, the mode associated with the controller (i.e. 8th) moves away from the imaginary axis. The voltage transducer modes (11th and 12th) are quite stable. The 13th and 14th modes are introduced with the advent of the circulating current controllers. Both have a damping ratio close to one. Though the eigenvalue of the 14th mode is somewhat close to the imaginary axis, it is significantly for distant and stable.

Fig. 3.5 shows the root-locus for Case B, with the varying in PI parameters. The proportional variables K_{pA} and K_{pB} of the PI controller are represented by K_p. Likewise, the integral parameters K_{iA} and K_{iB} are represented by K_i. K_{pcir} and K_{icir} are PI parameters for a circulating current controller. The system has demonstrated stability across a wide range of parameter variations. The values of K_p and K_i are varied from 5 to 2000 and 10 to 5000, respectively. As the value of K_p increases, the few poles near the imaginary axis move closer to zero. When K_i increases, these poles shift away from the imaginary axis. Similarly, K_{pcir} and K_{icir} are varied from 50 to 5000 and 10 to 10000, respectively. When K_{pcir} is increased,

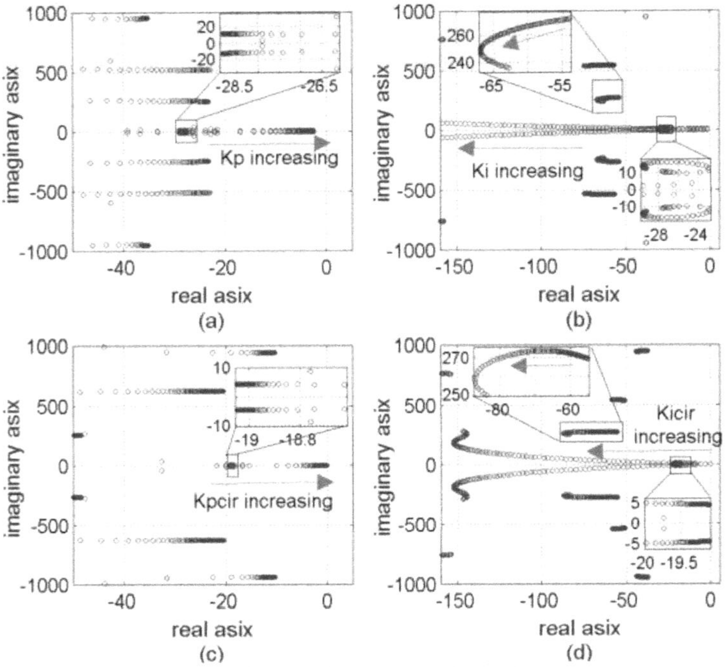

Fig. 3.5. Root locus for Case A (a) $5 < K_p < 2000$ **(b)** $10 < K_i < 5000$ **(c)** $50 < K_{pcir} < 5000$ **and (d)** $10 < K_{icir} < 10000$

the respective poles move toward the zero axis, and when K_{icir} is increased, the respective poles move away from the zero axis. Both the eigenvalues and the root-locus plots establish that the developed models are stable.

3.6 RTDS Simulation and Impact of Circulating Current Controller

The results of real-time simulations utilising RTDS are depicted in Figs. 3.6 and 3.7. Because all six arms react similarly, this section only displays the graphs for the upper arm of phase 'a'. These data show how MMCs operate with and without a circulating current controller. The operating point has been set to 480 MW and 0 MVAr. The ac and dc voltages are kept at rated.

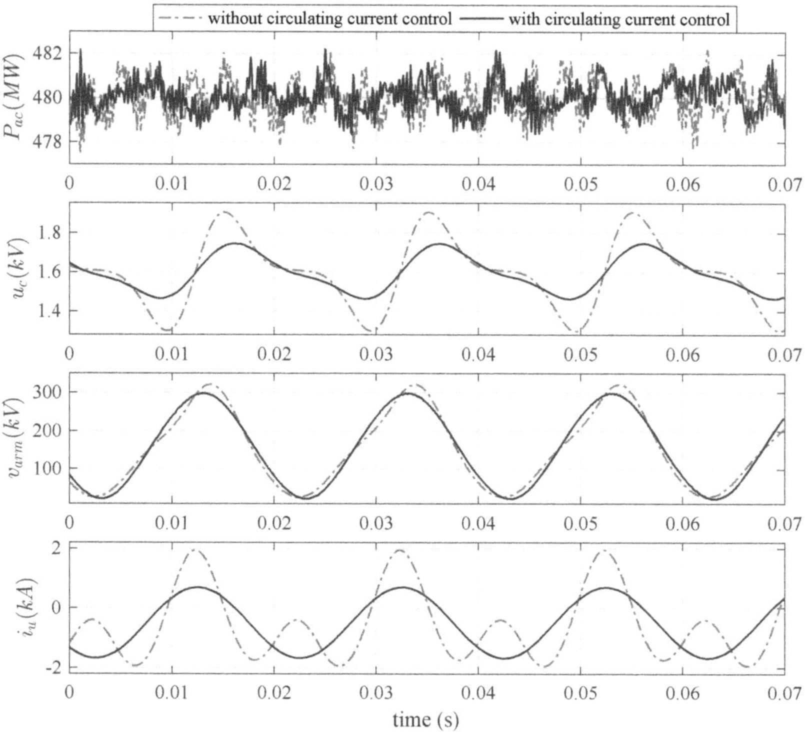

Fig. 3.6. (a) Active power, (b) SM capacitor voltage, (c) arm voltage, and (d) arm current

Fig. 3.6 illustrates a comparative steady-state analysis of the MMC operation with and without an auxiliary controller. The power flow is not affected by the presence of the circulating current controller. However, the internal dynamics of the MMC is significantly influenced. The SM capacitor voltage oscillates between 1.29 kV and 1.9 kV without an auxiliary controller. The lower and higher limits of the oscillation-band shift to 1.47 kV and 1.75 kV, respectively, when the auxiliary controller is active. The suppression of the second-order components leads to a substantially less distorted shifted sinusoidal waveform. The reduction of the SM capacitor voltage peak reduces the stress on the capacitors. In some situations, it may lead to a reduction in the system's capacitor requirements. The peak of arm voltage is also reduced from 321 kV to 297 kV, as shown by the v_{arm} plots. The arm

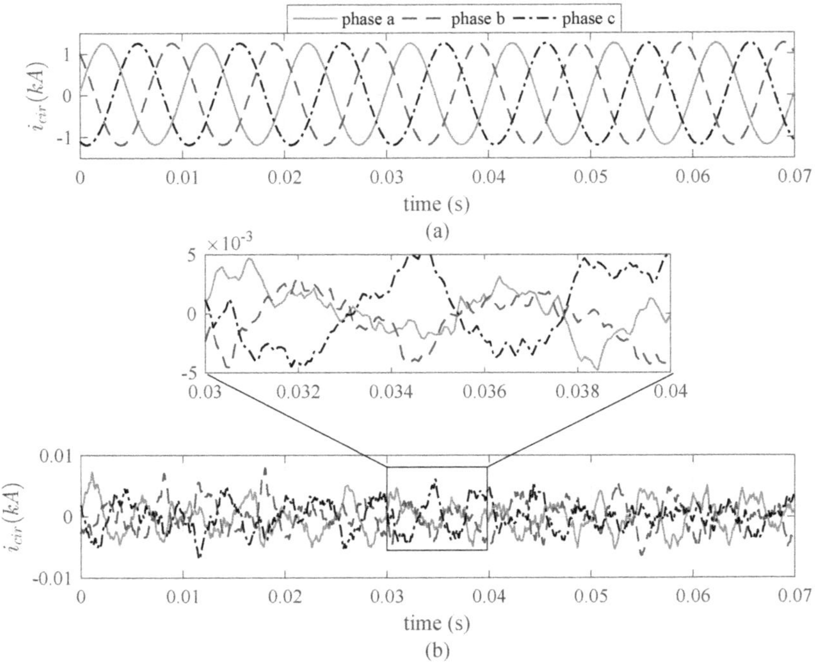

Fig. 3.7. Circulating current (a) without circulating current controller, and (b) with circulating current controller

current witnesses the most significant effect of the auxiliary controller. When MMC is run without the auxiliary controller, the arm current's positive and negative peak values are 1.94 kA and -1.94 kA, respectively. With the introduction of an auxiliary controller, the harmonic component is eliminated, and the peak values are reduced to 0.685 kA and -1.674 kA, respectively. The dc shift in arm current occurs due to the presence of dc current. Harmonics in arm currents are essentially the circulating currents illustrated in Fig. 3.7a, which have peak values of 1.255 kA and -1.185 kA, respectively. The auxiliary controller reduces second-order components, resulting in a 200-fold reduction in circulating currents (Fig. 3.7b).

Thus, using a circulating current controller minimises the voltage stress of SM capacitors while reducing the second-order harmonic component of the arm current. This, in turn, minimises the power loss as well as the heat sink requirement.

3.7 Conclusion

The modular structure of MMC has numerous advantages over conventional converter topologies. However, the modular approach complicates the modelling and operations. This chapter has successfully developed the small-signal and EMT models of the MMC. Two configurations are studied to assess the impact of circulating current with and without a circulating current controller. The small-signal models developed are of the 24th and 20th orders. Both the small-signal and EMT models demonstrated the same dynamic response, proving the study's validity. The controllers have been designed using the conventional method. Eigenvalue analysis reveals that most of the modes are substantially stable and damped. Only a few modes associated with the SM capacitor voltages are poorly damped. The addition of auxiliary control improves two of the poorly damped modes while having a marginal detrimental effect on two others. The variation of modes in relation to controller parameters is shown by the root locus, which proves the stability over a large range of parameter values. Furthermore, RTDS simulation demonstrates that employing an auxiliary controller reduces the circulating current by 200 folds and also reduces the SM capacitor voltage ripples. This results in less voltage stress on the capacitor, lower power losses, and lower cooling requirements.

Chapter 4

Design of MMC Controller Based on Internal Dynamics for HVDC

4.1 Introduction

MMC control, as outlined in Chapter 2, has a multifaceted challenge. Power flow and voltage regulation are two of the most important objectives. The internal dynamics of the MMC play a dominant role in its operation. However, the most common control approaches ignore this aspect. One such approach is discussed in Chapter 3. The presence of 3N capacitors leads to complexities that can not be ignored while designing the controller.

The authors of [65] have investigated the interaction of voltage and current within the SM. This interaction gives rise to the MMC's internal dynamics. A clear insight into the EMT model considering different dynamic equations has been provided in [58, 59]. The authors of [139] highlighted how harmonic currents could influence the operating range. Furthermore, as demonstrated in [60], as well as in Chapter 3, the stability is dependent on the dynamics of the SM capacitor. Internal dynamics should, therefore, be considered during the controller design process. Few model predictive control studies [69, 70] have been carried out, but their methods place an excessive computing burden on the system. In addition, the authors of [79] took into account the interaction of electrical variables in SM and the internal dynamics in MMC modelling and primary controller design. Their study, however, is limited to the second-order harmonic voltage ripple in SM capacitors. Moreover, they have overlooked the internal dynamics of the auxiliary controller design without giving any explanation and have used the conventional model.

In this chapter, considering the internal dynamics, a detailed model for the fundamental and circulating currents of the MMC has been developed. Three sets of controllers have been derived by considering the inter-couplings and cross-couplings of models. Compared to conventional control methods, these controllers enable improved power decoupling. The significance of inter-couplings, cross-couplings and SM capacitor voltage ripples in overall operation has been analysed. The comparative performance analysis of proposed controllers reveals that the proposed simplified version of the controller outperforms the other control methods.

4.2 Analysis of Conventional Model and Controller

The arm currents of the MMC are a combination of dc, fundamental ac and circulating current as defined in (3.3). As described in the previous chapter, the widely accepted method for controlling the MMC is to use two parallel controllers, one primary and one auxiliary, to control fundamental and circulating current. Fig. 4.1 depicts the equivalent circuits for fundamental and second order circulating current.

The mathematical model of the same can be defined as in (4.1) and (4.2), respectively, for fundamental and circulating current. These S-domain equations are equivalent to the rearrangement of the differential equations defined in (3.11) and

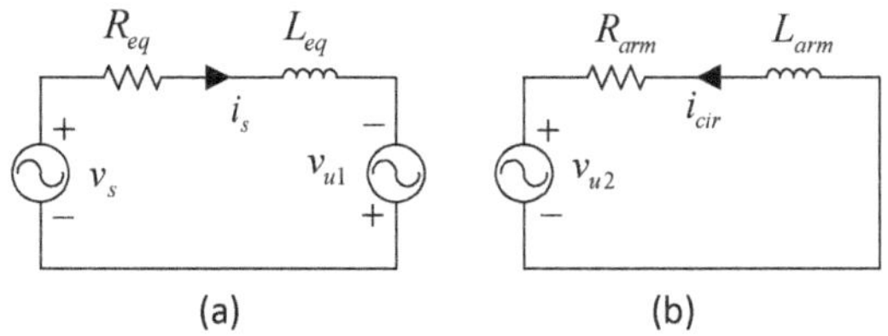

Fig. 4.1. MMC equivalent circuits for (a) fundamental frequency output current (b) 2nd order circulating current

(3.12).

$$v_{u1} = -V_s + (S * L_{eq} + R_{eq})I_s \tag{4.1}$$

$$v_{u2} = -(S * L_{arm} + R_{arm})I_{cir} \tag{4.2}$$

For this conventional circuit, the d-q model (also known as the control plant) of the fundamental ac and the 2nd order circulating current are presented in Figs. 4.2b and 4.2d, respectively. Conventionally, the controllers have been designed using these models as reference. Figs. 4.2a and 4.2c depict d-q based conventional primary and auxiliary controllers, respectively. The primary controller is derived considering the d-q model of fundamental ac voltage and current. The auxiliary control is based on the 2nd order circulating the current model.

Since these models and the respective controllers are based on the traditional understanding of the converter, they overlook the internal dynamics of MMC. Traditionally, the capacitor is connected to the dc terminal, isolated from ac network,

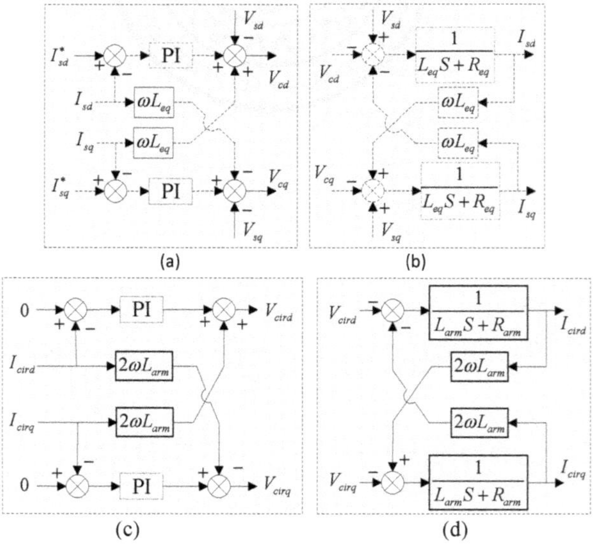

Fig. 4.2. d-q based conventional (a) primary controller (b) fundamental frequency ac current model (c) auxiliary controller (d) 2nd order circulating current model

while in MMC, the SM capacitors are interlinked with both ac and dc networks. The transformation from a three-phase model to a d-q model replaces the alternating variable with equivalent d-q-axis variables which are constant [66, 140]. It simplifies system control since tracking a constant reference is significantly easier than tracking an alternating reference, which suffers from phase lag. The d-q equivalents of the three-phase ac systems have the cross-couplings between the d- and q-axis. It is caused due to the presence of the inductor. As shown in Fig. 4.2b, in conventional VSCs, the d-q-axis variables are coupled together by an inductor L_{eq}. The q-axis coupling component ($\omega L_{eq} I_{sq}$) is subtracted from the input voltage of the d-axis, and similarly, the d-axis coupling component ($\omega L_{eq} I_{sd}$) is added in the input voltage of q-axis. Thus, the change in the d-axis variable affects the q-axis variable and vice versa. This coupling is reflected as a coupling between active (P_{ac}) and reactive power (Q_{ac}).

Similar coupling components are included in the controllers to decouple the power components. In the d-q-based PI controller depicted in Fig. 4.2a, the q-axis coupling component ($\omega L_{eq} I_{sq}$) is added to the d-axis controller output, and the d-axis coupling component ($\omega L_{eq} I_{sd}$) is subtracted from the d-axis controller output. Decoupling enables independent control of the d- and q-axis currents, and thus of P_{ac} and Q_{ac}. Similarly, in the circulating current model of Fig. 4.2d, the d and q components are coupled through the arm inductor L_{arm}. The coupling component $2\omega L_{arm} I_{cird}$ is added to the q-axis. Likewise, $2\omega L_{arm} I_{cirq}$ is added to the d-axis. To facilitate the decoupling, in the controller of Fig. 4.2c, $2\omega L_{arm} I_{cirq}$ is added to d-axis and $2\omega L_{arm} I_{cird}$ is subtracted from the q-axis.

The decoupling results in independent control of d- and q-axis current and thus of P_{ac} and Q_{ac}. Usually, d-q based PI controller successfully facilitates the power decoupling. However, the same is not evident in MMC due to the increased coupling between electrical quantities. Furthermore, the controller's performance is not as good as desired. The following section provides a detailed study of the same.

4.3 Detailed Modelling of MMC

The internal dynamics of the MMC are taken into account when developing an efficient controller. Authors in [60, 141] have provided insight into the same. The primary equations of MMC operation are defined in Section 3.2.

Substituting average switching function (3.4) and SM capacitor voltage (3.6) in (3.1) gives the detailed expression for upper arm voltage as (4.3), where v_{udc} is the dc component and v_{uk} is the kth order order component of upper arm voltage, as defined in (4.4). The expression for lower arm voltage is similar to that of upper arm voltage, with identical dc and even order components. However, the odd-order components are of opposite polarity.

$$v_u = v_{udc} + v_{u1} + v_{u2} + v_{u3} + v_{u4} + v_{u5} \tag{4.3}$$

$$\begin{aligned}
v_{udc} =& \frac{N}{2}\overline{u}_c - \frac{N}{4}Mu_{c1}\cos(\beta_f - \theta_1) + \frac{Nv_{cir}}{2V_{dc}}u_{c2}\cos(\beta_c - \theta_2)\\
v_{u1} =& -\frac{N}{2}M\overline{u}_c\sin(\omega t + \beta_f) - \frac{N}{4}Mu_{c2}\cos(\omega t + \theta_2 - \beta_f) + \frac{N}{2}u_{c1}\sin(\omega t + \theta_1)\\
&+ \frac{Nv_{cir}}{2V_{dc}}u_{c1}\cos(\omega t + \beta_c - \theta_1) + \frac{Nv_{cir}}{2V_{dc}}u_{c3}\cos(\omega t + \theta_3 - \beta_c)\\
v_{u2} =& \frac{Nv_{cir}}{V_{dc}}\overline{u}_c\sin(2\omega t + \beta_c) + \frac{N}{4}Mu_{c1}\cos(2\omega t + \beta_f + \theta_1) + \frac{N}{2}u_{c2}\sin(2\omega t + \theta_2)\\
&- \frac{N}{4}Mu_{c3}\cos(2\omega t + \theta_3 - \beta_f)\\
v_{u3} =& -\frac{Nv_{cir}}{2V_{dc}}u_{c1}\cos(3\omega t + \beta_c + \theta_1) + \frac{N}{2}u_{c3}\sin(3\omega t + \theta_3)\\
&+ \frac{N}{4}Mu_{c2}\cos(3\omega t + \theta_2 + \beta_f)\\
v_{u4} =& -\frac{Nv_{cir}}{2V_{dc}}u_{c2}\cos(4\omega t + \beta_c + \theta_2) + \frac{N}{4}Mu_{c3}\cos(4\omega t + \theta_3 + \beta_f)\\
v_{u5} =& -\frac{Nv_{cir}}{2V_{dc}}u_{c3}\cos(5\omega t + \theta_3 + \beta_c)
\end{aligned} \tag{4.4}$$

By applying KVL, the fundamental frequency and second-order circulating current, respectively specified in (3.11) and (3.12), are obtained. Substituting values from (4.4) to (3.11) and (3.12) and applying the transformation using (3.8), results in the detailed expressions of fundamental and circulating currents in d-q coordinates as in (4.5) and (4.6), respectively. These equations take internal dynamics into account.

$$\begin{aligned}\frac{dI_{sd}}{dt} =& -\omega I_{sq} + \frac{V_{sd}}{L_{eq}} - \frac{R_{eq}I_{sd}}{L_{eq}} - \frac{NV_{cd}\overline{u}_c}{L_{eq}V_{dc}} + \frac{Nu_{c1d}}{2L_{eq}} + \frac{NV_{cirq}u_{c1q}}{2L_{eq}V_{dc}} + \frac{NV_{cird}u_{c1d}}{2L_{eq}V_{dc}} \\ &- \frac{NV_{cq}u_{c2q}}{2L_{eq}V_{dc}} - \frac{NV_{cd}u_{c2d}}{2L_{eq}V_{dc}} + \frac{NV_{cirq}u_{c3x}}{2L_{eq}V_{dc}} + \frac{NV_{cird}u_{c3y}}{2L_{eq}V_{dc}} \\ \frac{dI_{sq}}{dt} =& \omega I_{sd} + \frac{V_{sq}}{L_{eq}} - \frac{R_{eq}I_{sq}}{L_{eq}} - \frac{NV_{cq}\overline{u}_c}{L_{eq}V_{dc}} + \frac{Nu_{c1q}}{2L_{eq}} + \frac{NV_{cird}u_{c1q}}{2L_{eq}V_{dc}} + \frac{NV_{cirq}u_{c1d}}{2L_{eq}V_{dc}} \\ &- \frac{NV_{cq}u_{c2d}}{2L_{eq}V_{dc}} - \frac{NV_{cd}u_{c2q}}{2L_{eq}V_{dc}} - \frac{NV_{cird}u_{c3x}}{2L_{eq}V_{dc}} - \frac{NV_{cirq}u_{c3y}}{2L_{eq}V_{dc}}\end{aligned} \tag{4.5}$$

$$\begin{aligned}\frac{dI_{cird}}{dt} =& -2\omega I_{cirq} - \frac{R_{arm}}{L_{arm}}I_{cird} - \frac{NV_{cird}\overline{u}_c}{L_{arm}V_{dc}} - \frac{NV_{cq}u_{c1q}}{2L_{arm}V_{dc}} + \frac{NV_{cd}u_{c1d}}{2L_{arm}V_{dc}} - \frac{Nu_{c2d}}{2L_{arm}} \\ &+ \frac{NV_{cq}u_{c3x}}{2L_{arm}V_{dc}} + \frac{NV_{cd}u_{c3y}}{2L_{arm}V_{dc}} \\ \frac{dI_{cirq}}{dt} =& +2\omega I_{cird} - \frac{R_{arm}}{L_{arm}}I_{cirq} - \frac{NV_{cirq}\overline{u}_c}{L_{arm}V_{dc}} - \frac{NV_{cq}u_{c1d}}{2L_{arm}V_{dc}} + \frac{NV_{cd}u_{c1q}}{2L_{arm}V_{dc}} - \frac{Nu_{c2q}}{2L_{arm}} \\ &+ \frac{NV_{cq}u_{c3y}}{2L_{arm}V_{dc}} + \frac{NV_{cd}u_{c3x}}{2L_{arm}V_{dc}}\end{aligned} \tag{4.6}$$

where d-q variables can be defined as $I_{sd} = I_s \sin\alpha_s$, $I_{sq} = I_s \cos\alpha_s$, $V_{sd} = V_s \sin(\alpha)$, $V_{sq} = V_s \cos(\alpha)$, $I_{cird} = I_{cir} \sin\alpha_c$, $I_{cirq} = I_{cir}\cos\alpha_c$, $u_{c1d} = u_{c1}\sin\theta_1$, $u_{c1q} = u_{c1}\cos\theta_1$, $u_{c2d} = u_{c2}\sin\theta_2$, $u_{c2q} = u_{c2}\cos\theta_2$, $u_{c3x} = u_{c3}\cos\theta_3$, $u_{c3y} = u_{c3}\sin\theta_3$, $V_{cd} = \frac{1}{2}MV_{dc}\sin\beta_f$, $V_{cq} = \frac{1}{2}MV_{dc}\cos\beta_f$, $V_{cird} = v_{cir}\sin\beta$, $V_{cirq} = v_{cir}\cos\beta_c$

Using (4.5) and (4.6), detailed d-q models for the fundamental and circulating current of MMC have been developed as depicted in Figs. 4.3a and 4.3b, respectively. These models take into account voltage harmonic components up to the third-order. As mentioned earlier, 3N capacitors of MMC are distributed among its six arms. These capacitors facilitate the energy exchange between the ac and dc sides of MMC. Thus, the fundamental current I_{sd}, I_{sq}, and the dc side current I_{dc} are mutually coupled through the SM capacitors. This results in a coupling between active power P_{ac}, reactive power Q_{ac}, and dc power P_{dc}. Further, as expressed by

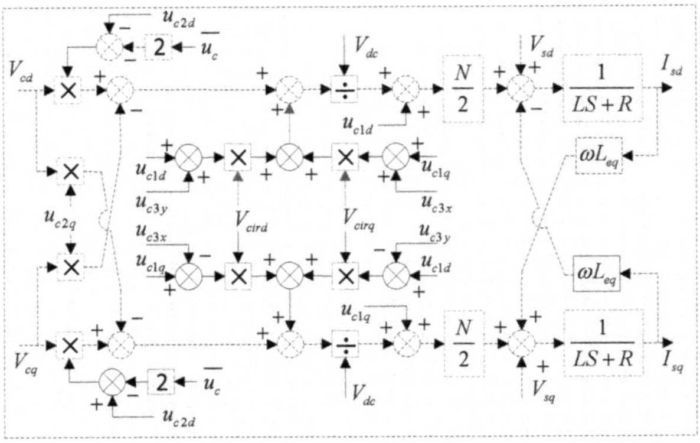

(a)

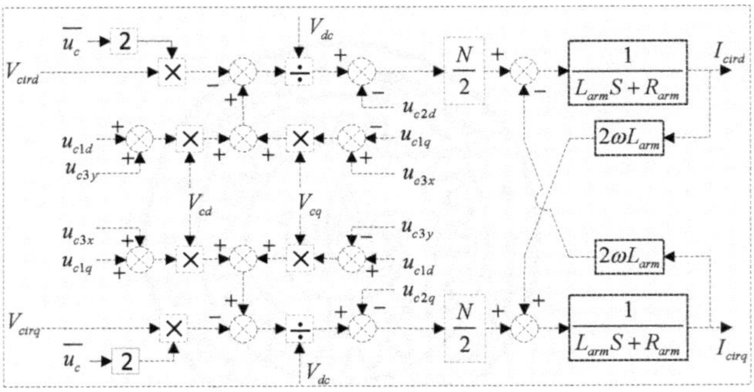

(b)

Fig. 4.3. Detailed d-q model of (a) fundamental frequency output current of MMC (b) 2nd order circulating current of MMC

(3.1), (4.3), and (4.4), the control signals for maintaining the desired fundamental and dc currents, also generate a 2nd- and higher-order frequency voltages, which in-turns induce undesired circulating currents. These undesirable circulating current factors are inextricably linked to dc and fundamental ac variables. These coupled three-phase variables are transformed into d-q coordinates using dynamic phasors. As is evident from Fig. 4.3a and (4.5), the d-q-axis variables of fundamental frequency current model are coupled through the inductor (L_{eq}) as well as the capacitor voltage component (u_{c2q}). Similarly, as illustrated in Fig. 4.3b and (4.6), the d-q variables of the second-order harmonic circulating current are coupled through the inductor (L_{arm}).

Furthermore, the inputs to the fundamental frequency current model (V_{cd} and V_{cq}) and the circulating current model (V_{cird} and V_{cirq}) interacts with SM capacitor voltage. The resultant is added to the circulating and fundamental current models, respectively. To summarise, the detailed fundamental frequency current model contains two cross-couplings, whereas the traditional model of Fig. 4.2 has only one. Similar to the conventional model, the d and q axes are cross-coupled via the inductor L_{eq}. An additional coupling is caused by 2nd order harmonic voltage ripple of SM capacitor. Further, there exists an inter-coupling between fundamental and circulating current models. Both are coupled through SM capacitor voltage ripple. The fundamental and second-order circulating voltages (V_{cd}, V_{cq}, V_{cird} and V_{cirq}) interact with the capacitor voltage ripples, and their outcome forms the inter-coupling component.

4.4 Proposed Internal Dynamics Based Controllers

On the basis of the developed models, new sets of controllers have been derived. These controllers are 1) detailed controller, 2) mixed controller and 3) simplified controller. In the design of a conventional controller (Fig. 4.2a), the coupling effect of the inductor on d-q-axis variables of the current model is neutralised by adding/subtracting the respective component in the controller. A similar philosophy has been applied to develop the proposed controllers. Additionally, the impact analysis of coupling components aids in simplifying the control structure.

4.4.1 Detailed controller

The detailed model of MMC includes cross-couplings and inter-couplings, which were not considered in conventional models. In addition, the SM capacitor voltage ripple up to 3rd order has been included. Detailed primary and auxiliary controllers have been designed with these factors in mind, as shown in Figs. 4.4a and 4.4b, respectively. When these two controllers operate in parallel, the combined structure is referred to as the detailed controller.

The coupling effect of inductor L_{eq} is expressed by subtraction of $\omega L_{eq} I_{sq}$ and addition of $\omega L_{eq} I_{sd}$, respectively, in d- and q-axis of the fundamental frequency current model (in Fig. 4.3a). Similarly, in the 2nd order circulating current model (Fig. 4.3b), the influence of inductor L_{arm} is represented by subtracting $2\omega L_{arm} I_{cirq}$ and adding $2\omega L_{arm} I_{cird}$ in the d and q axes, respectively. An equivalent term is

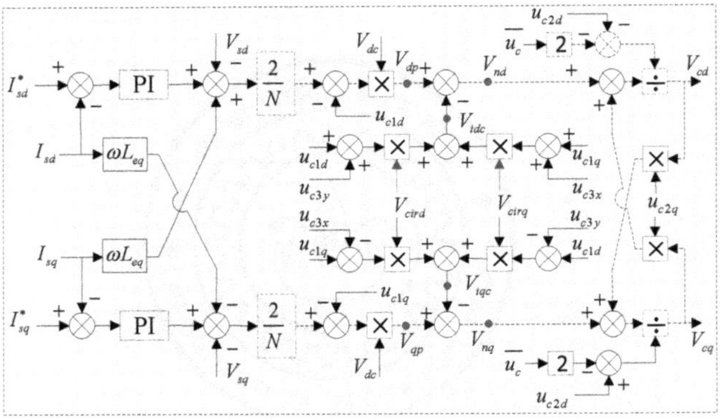

(a)

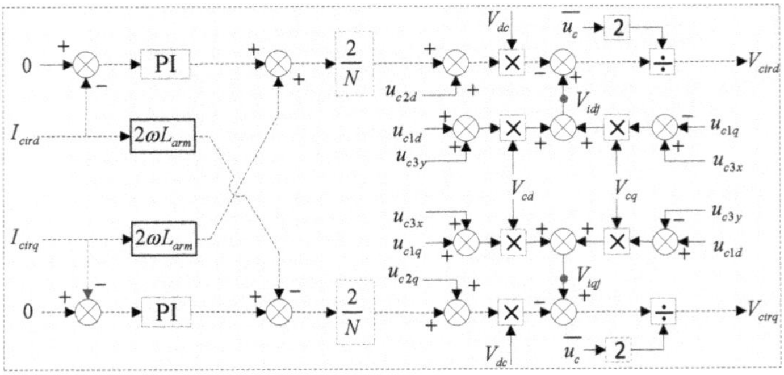

(b)

Fig. 4.4. (a) Detailed primary controller (b) Detailed auxiliary controller

introduced in the detailed controller to cancel out the inductor's coupling effect. In Fig. 4.4a, $\omega L_{eq} I_{sq}$ is added, whereas $\omega L_{eq} I_{sd}$ is subtracted from the PI controller output of d- and q-axis, respectively. Similarly, in Fig. 4.4b, $2\omega L_{arm} I_{cirq}$ is added to the d-axis and $2\omega L_{arm} I_{cird}$ is subtracted from the q-axis. The additional cross-coupling in the fundamental frequency current model is from 2nd harmonic SM capacitor voltage component u_{2cq}. As illustrated in Fig. 4.3a, $u_{2cq} V_{cq}$ is subtracted from d-axis variable $-(2\overline{u}_c + u_{2cd})V_{cd}$, and $u_{2cq} V_{cd}$ is subtracted from q-axis variable $-(2\overline{u}_c - u_{2cd})V_{cq}$. To nullify this cross-coupling effect in Fig. 4.4a, $u_{2cq} V_{cq}$ and $u_{2cq} V_{cd}$ are added in V_{nd} and V_{nq}, respectively. As mentioned in the previous section, there also exists an inter-coupling between the fundamental and circulating current model. The functions of circulating current reference voltages V_{cird} and V_{cirq}, and voltages ripples are added in d- and q-axis of Fig. 4.3a. To nullify the effect of these inter-coupling components, the variables V_{idc} and V_{iqc} are included in the primary controller of Fig. 4.4a. V_{idc} and V_{iqc} are functions of circulating current controller output V_{cird}, V_{cirq}, and voltage ripples. Similarly, in the case of the circulating current model of Fig. 4.3b, the inter-coupling is a function of V_{cd}, V_{cq}, and voltage ripples. To nullify them, variable V_{idf} and V_{iqf} are included the circulating current controller, shown in Fig. 4.4b. These variables are a function of primary controller outputs V_{cd}, and V_{cq}, and voltage ripples.

The proposed detailed controller is intended to reduce cross-couplings caused by inductor L_{eq}, L_{arm}, and the SM capacitor voltage component u_{c2q}, as well as inter-couplings between the fundamental and circulating current models. The detailed controller experiences increased circulating current and ripples in dc. Therein, the magnitude of the inter-coupling signal from the primary to the auxiliary controller is significantly larger (approx 10x) than the signal of the auxiliary controller. Even a small change for the primary controller is significantly large with respect to the auxiliary controller. Therefore, inter-coupling from the primary to the auxiliary controller has a detrimental effect on the auxiliary controller's performance.

4.4.2 Mixed controller

To overcome the shortcomings of the detailed controller while minimising complexity, a combination of the detailed primary controller (Fig. 4.4a) and the conventional auxiliary controller (Fig. 4.2d) is being used. The new arrangement is referred to as a "mixed controller." This controller takes into account the d-q-axis cross-coupling as well as the inter-coupling from the circulating to the fundamental current model. The inter-coupling from the fundamental to the circulating current model, on the other hand, has been overlooked.

4.4.3 Simplified controller

Subtraction of the inter-coupling component V_{idc} from V_{dp} yields V_{nd}, as shown in Fig. 4.4a. Similarly, on the q-axis, the combination of V_{iqc} and V_{qp} yields V_{nq}. The final control command is derived using these two signals, V_{nd} and V_{nq}. The controller's inter-coupling has been analysed in order to further reduce complexity. The system under consideration is initially operating at 400 MW and 0 MVAr. The active power is raised by 50 MW at time instant of 0.04 seconds. Fig. 4.5 presents the ratio of the d-axis and q-axis components, respectively. d_{ratio} and q_{ratio} are defined as in (4.7). Initially, under steady-state, the inter-coupling component V_{idc} is 2% of Vdp and on

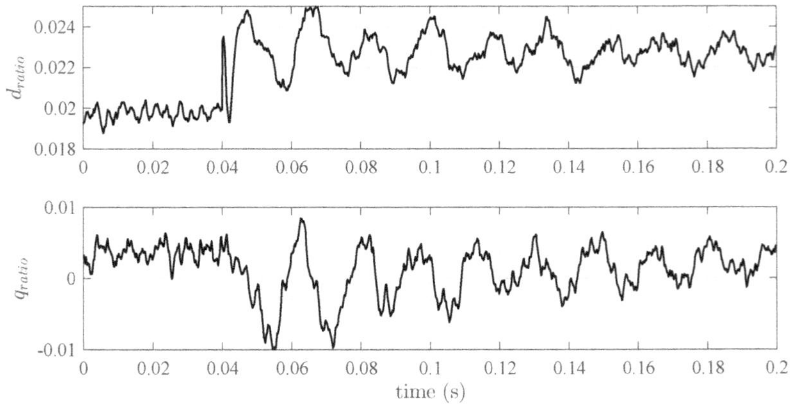

Fig. 4.5. Coupling analysis of mixed controller

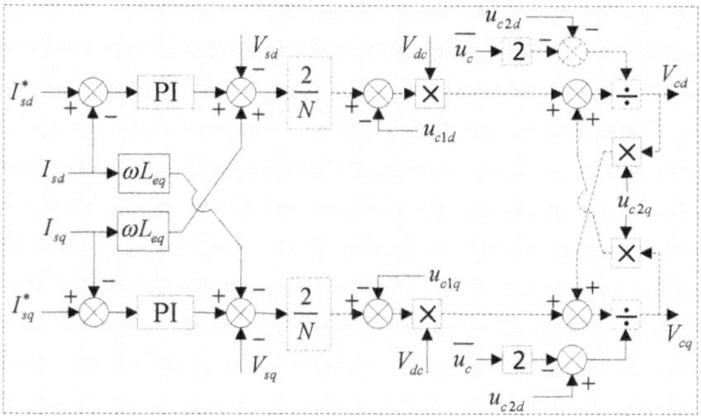

Fig. 4.6. Simplified primary controller

q-axis V_{iqc} is less than 1% of V_{qp}. Further, under the transient condition, when the active power has been raised by 50 MW, the ratios attain maximum values of 2.5% and 1%, respectively.

$$d_{ratio} = \frac{V_{idc}}{V_{dp}}, \qquad q_{ratio} = \frac{V_{iqc}}{V_{qp}} \tag{4.7}$$

Since the inter-coupling components are relatively small and have no substantial impact on the overall performance of the controller, they have been eliminated to create the simpler structure shown in Fig. 4.6. For the circulating current control, the conventional auxiliary controller has been incorporated. The combined set is referred to as a "simplified controller," as it takes into account the additional cross-coupling between the d-q-axis while ignoring the inter-couplings entirely.

4.5 Results and Discussion

The controllers' performance has been evaluated using a real-time digital simulator (RTDS). The MMCs under investigation are rated at 500 MVA, 166 kV line-to-line ac voltage, and 320 kV pole-to-pole dc voltage. The four MMCs, which are identical in all respect except the controller, have been designed and operated simultaneously. The four controllers (conventional, detailed, mixed, and simplified) all have identical PI parameter values. The proportional and integral constants (K_p and K_i) of the primary controller are set to 75 and 500, respectively. The same parameters for the auxiliary controller are set to 60 and 1000, respectively. The performance comparison of the controllers is shown in Fig. 4.7, Fig. 4.8, and Fig. 4.9. Dashed lines represent plots for the conventional controller. The detailed controller is represented by dash-dot lines, whereas the simplified controller is represented by solid lines. Plots for mix controllers are omitted since they overlap with plots for the simplified controller. Three case scenarios have been selected for analysis:

1) the base case

2) change in active power

3) change in reactive power

4.5.1 Base case operation

The base case is primarily a performance analysis of the controllers under steady-state conditions. The operating conditions are set to 400 MW and 0 MVAr, with rated ac and dc terminal voltages. Fig. 4.7 represents the base case operation. The plots show that three controllers effectively maintain the reference active (P_{ac}) and reactive power (Q_{ac}), as well as the rated dc voltage (V_{dc}), steady. The I_{dc} represents the current flow over the dc line. With constant dc voltage, this also provides a snapshot of dc power flow. With the detailed controller, I_{dc} has oscillating components, which cause dc power oscillation. The conventional, as well as simplified controller, does not experience this drawback. I_{cird} and I_{cirq} are d-q components of circulation current. Their plots demonstrate that all three controller setups successfully eliminate the second-order circulating current.

The plot for $\overline{u}_c$ represents the dc component of the SM capacitor voltage. However, as shown in the plots, it exhibits oscillations as well. The oscillations are rather large in the case of the detailed controller, whereas they are substantially smaller in the case of the simplified and conventional controllers. Furthermore, the average dc voltage from the simplified controller is slightly lower than that of the conventional

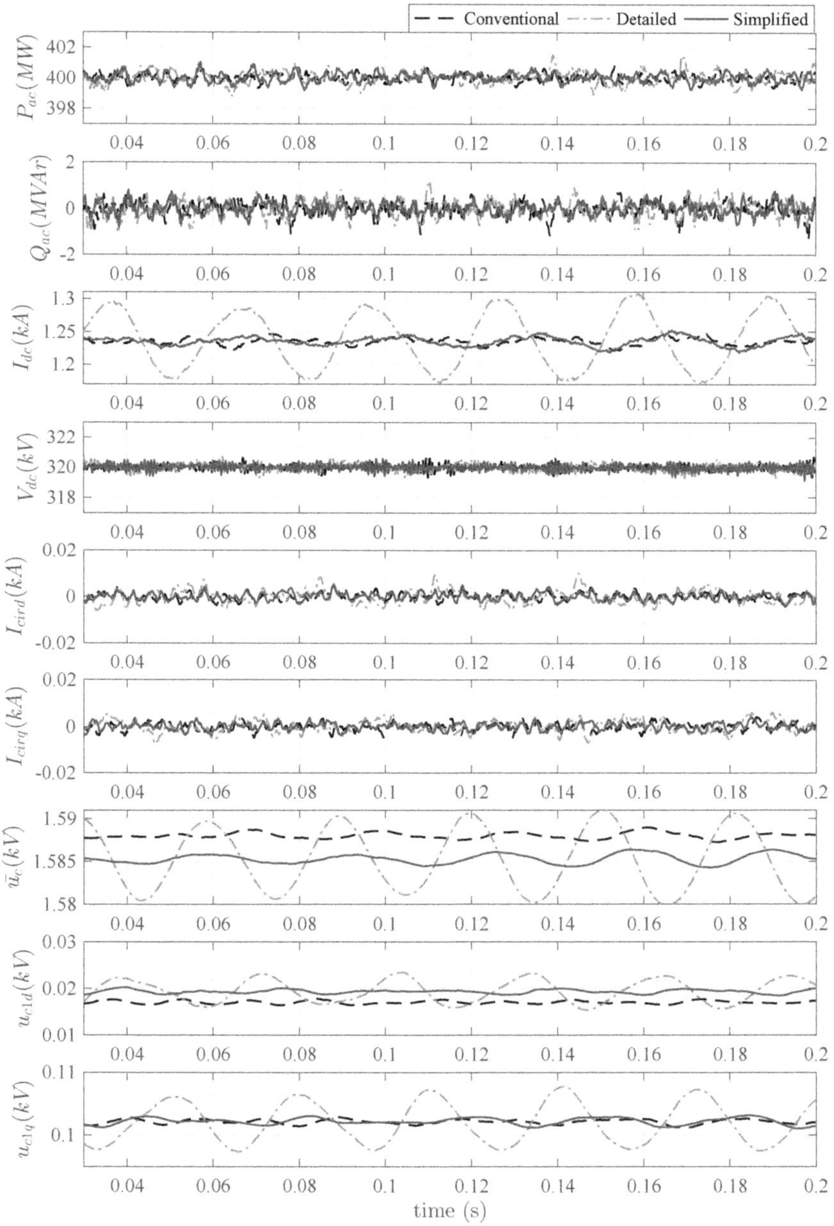

Fig. 4.7. Result for controller performance for base case with $P_{ac} = 400$ MW and $Q_{ac} = 0$ MVAr

controller, which can lessen the stress on the SM capacitor. In terms of fundamental frequency ripple of SM capacitor voltage (i.e. u_{c1d} and u_{c1q}), the detailed controller performs poorly. The simplified controller, on the other hand, delivers a slightly greater d-component value than conventional controllers. The conventional and simplified controllers yield nearly constant values for u_{c1d}, with the simplified controller delivering a little higher value. However, the detailed controller leads to sustained oscillations. Likewise, for u_{c1q}, the detailed controller exhibits prolonged oscillations, whereas the simplified and conventional controllers produce values that are almost steady and overlapping.

In summary, for steady-state operation, the performance of the simplified controller is as good as that of the conventional controller, even slightly better in terms of SM capacitor voltage. However, the detailed controller performs poorly because it introduces ripples in dc current, dc power, and SM capacitor voltage. In terms of internal dynamics, the simplified controller performs equivalent to the conventional controller.

4.5.2 Change in active power

The transient case scenario for the controller performance comparison is depicted in Fig. 4.8. Initially, the MMCs are operating at base case values (P_{ac} = 400 MW, Q_{ac} = 0 MVAr, and rated voltages). At the time instant of 0.04 second, the active power reference is increased by 50 MW. All the controllers responded relatively similar in terms of active power flow. It obtains the new reference almost instantly and swiftly settles on the final value. The overshoot is maximum for the detailed controller (i.e. 14 MW) and minimum for the conventional one (i.e. 7 MW). For the simplified controller, it is 11.5 MW, which is in between the values for the other two controllers. However, the simplified controller attains the almost stable reference value slightly earlier than the conventional controller. In comparison, the detailed controller takes the longest and results in prolonged oscillation, albeit of small magnitude.

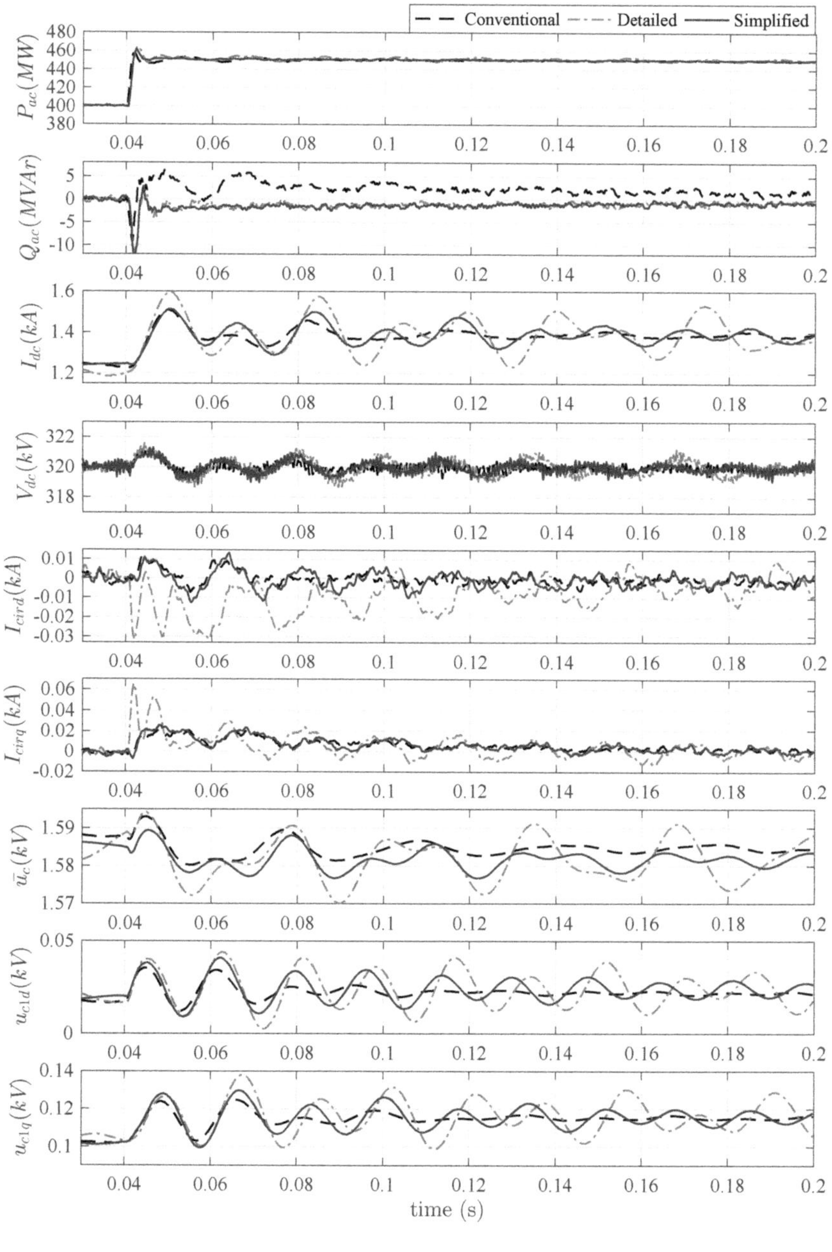

Fig. 4.8. Result for controller performance for change in active power from 400MW to 450MW

The impact on reactive power is significantly different. In the instant of change, all three controllers lead to overshoot, as in the case of active power flow. However, with proposed controllers, the reactive power exchange drops to near-zero in 15-20 milliseconds and eventually reaches zero. In the case of a conventional one, the reactive power exchange sustains significantly higher values for a much longer period of time. In other words, the proposed controllers outperform the conventional controller in terms of decoupling. The dc current plots show a similar observation in the transient scenario as it did in the base case. The detailed controller has higher overshoot and oscillations, whereas the conventional controller has lower overshoot values and achieves a steady state at a much faster rate. At the instant of transient, the simplified controller traces the path of the conventional controller but takes a few extra milliseconds to settle. The magnitude and time period of the first oscillation for the simplified controller are identical to those of the conventional controller. However, in successive cycles, the time-period of the oscillations increases gradually for the simplified controller compared to the conventional controller. In addition, the simplified controller has a slower decay of oscillation amplitude than the conventional controller. The detailed controller exhibits the highest magnitude of oscillation and takes the longest time to achieve the steady-state value.

The terminal dc voltage experiences a sudden change in values and oscillations, as a result of power redistribution among the SM capacitor. The very low magnitude oscillation sustains for a while and dies out over time. In this case, too, the detailed controller has the largest magnitude of oscillation in comparison to the other two controllers. The power redistribution increases the circulating current between the arms, which is reflected by I_{cird} and I_{cirq}. These variables attain a relatively high value with a detailed controller and take significantly longer to eliminate than the other two. As presented in I_{cird} plot, the conventional and simplified controller track the same path up to second cycles. Thereafter, the conventional controller attains a relatively constant value while the simplified controller exhibits small magnitude oscillations for a few more cycles prior to attaining a steady-state band. The I_{cirq}

plots, for both conventional and simplified controllers, trace the same path and settles in equal duration.

In terms of ripples in the dc component of SM capacitor voltage, the performance of the simplified controller is intermediate. While the magnitude of the oscillation is slightly greater for the simplified controller than for the conventional controller, it is substantially lower for the detailed controller. In terms of average values, it outperforms both counterparts while producing slightly lower values than the other two. Thus, reducing the voltage strain of SM capacitor more efficiently than its alternative. Similarly, for the fundamental components (u_{c1d} and u_{c1q}), the simplified controller's values are better than the detailed controller but slightly worse than the conventional controller. The conventional controller attains the relatively constant value in four oscillations. The signals from all the three controllers overlap for the 1st oscillations. However, the time-period for simplified and detailed controllers increases gradually. The magnitude of oscillation is smaller for the simplified controller than for the detailed controller, and it decays at a considerably faster rate. Thus, under transient conditions, the simplified controller provides better decoupling than the conventional controller while avoiding excessive ripples in variables, as in the case of the detailed controller.

4.5.3 Change in reactive power

Another transient scenario test is carried out by varying the reactive power support from MMC to the ac grid. The plots for the controller performance are shown in Fig. 4.9. The MMCs are reset to their original operating conditions with 400 MW active power flow and 0 MVAr reactive power support. At 0.04 seconds, the reactive power support from MMCs has been increased to 50 MVAr. As shown in reactive power plots, all three controllers efficiently achieve their primary objective of changing Q_{ac} support. As illustrated, the conventional controller has an overshoot of 7.5 MVAr, and the other two controllers have an overshoot of 8.8 MVAr. Since the difference is just 1.3 MVAr for a change of 50 MVAr, no controller has a

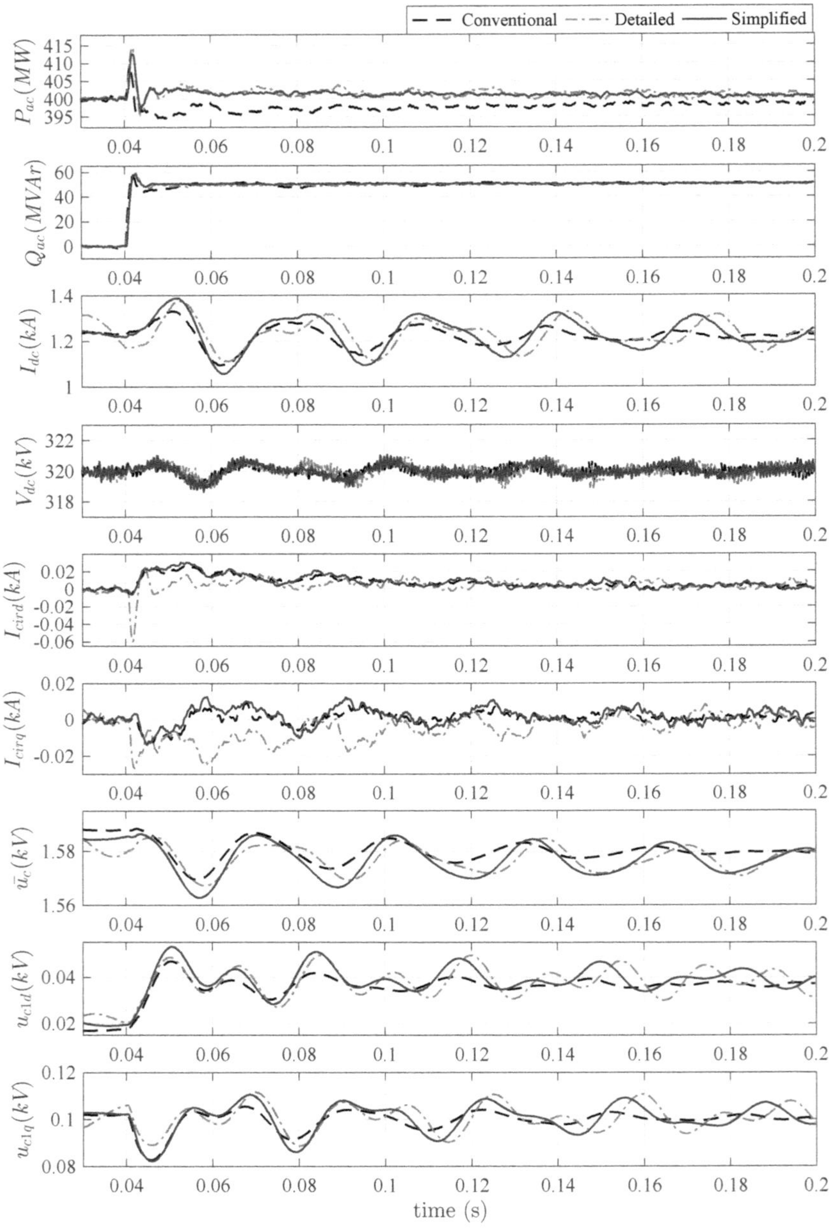

Fig. 4.9. Result for controller performance for change in reactive power from 0MVAr to 50MVAr

significant advantage over the others in terms of overshoot. In terms of attaining the constant reference value, the simplified controller outperforms the other two, as it settles earliest. Furthermore, as seen in the preceding test scenario, the simplified controller results in better decoupling, as evidenced by active power plots. When the Q_{ac} reference changes, all three controllers see a sudden increase in active power. Nonetheless, the simplified controllers re-attains a value in the ± 2 MW band across a reference of 400 MW in 20 milliseconds, whereas the other two controllers require substantially longer time periods and oscillations.

In terms of the dc current and terminal voltage, the performance of all three controllers are more or less the same. All three cause ripples in dc current, which settle the earliest in the case of a conventional controller, whereas simplified and detailed controllers have a larger magnitude and a slower decay rate of oscillation. The dc voltage ripples are negligible for all three MMCs. Re-distribution of power among SM capacitors causes the circulating current, as is depicted by the respective plots. For detailed controllers, the magnitudes of ripples are larger. The d-axis circulating current witnesses a high magnitude spike for the detailed controller at the time of transient, followed by gradually decreasing oscillations. The I_{cird} plots for the simplified and conventional controllers follow the same path and are free from spikes. For q-axis circulating current, all three controllers produce oscillations of relatively small size that eventually die out. The time it takes to eradicate them is almost the same for all three controllers. The simplified controller, which performs substantially better in terms of reducing SM capacitor voltage stress in the previous two test situations, does not provide a significant advantage in the event of reactive power change transients. All three controllers cause oscillation in the dc as well as the fundamental components of the SM capacitor voltage. The average voltage parameter values are similar for all three MMCs. Conventional controller plots achieve a relatively steady value faster than others. The magnitude of oscillation is almost equivalent for simplified and detailed controllers.

It can be summarised that the simplified controller facilitates better decoupling than the other two choices during the transient caused by a change in reactive power. In terms of circulating current, its performance is superior to that of a detailed controller but comparable to that of a conventional controller.

4.5.4 Discussion

The conventional MMC model ignores inter-coupling and cross-coupling. As a result, controllers based on conventional models are intrinsically incapable of dealing with increased coupling. The proposed detailed controller includes additional cross-couplings (between the d-q axis) and inter-couplings (between primary and auxiliary controller) components. Consideration of additional d-q-axis cross-coupling improves decoupling. On the contrary, considering inter-coupling leads to higher circulating current and dc current ripples. Further, the inter-coupling signals from the primary controller affect the auxiliary controller adversely. As even a minor change for the primary controller is significantly large for the auxiliary one. This disadvantage has been eliminated in the mixed controller.

In the mixed controller, the detailed auxiliary controller is replaced by the conventional auxiliary controller. Thus, it essentially ignores the inter-coupling from the fundamental current model to the circulating current model while retaining the inter-coupling from the auxiliary to the primary controller. Furthermore, as shown in Fig. 4.5, the contributions of inter-couplings of the auxiliary and primary controllers are insignificant and hence have been omitted to provide a simplified version. Exclusion of the primary-to-auxiliary inter-coupling decreases oscillation, whereas excluding the auxiliary-to-primary controller reduces complexity. Thus, the final version only considers cross-coupling, which is simpler than the detailed controller.

In terms of decoupling (active and reactive power), detailed and simpler controllers outperform conventional controllers. However, as previously stated, the detailed controller suffers from higher circulating current and dc current ripples. The simplified controller, on the other hand, avoids these problems. As a result, the simplified controller outperforms all others. However, a relatively higher magnitude and a slow rate of decay of oscillations in the arm variable can be seen as limitations of the simplified controller in comparison to the conventional controller. Moreover, the approach provided, here, can be extended to account for unbalanced grid situations.

4.6 Conclusion

MMC's modular structure has numerous advantages over other converter topologies. However, the modular approach complicates modelling and operations. Since MMC is a converter topology, the most common approach is to consider the conventional converter model and design the controller accordingly. The same concept has been put into practice in designing the primary and auxiliary controllers of the MMC.

This study has successfully developed a detailed d-q model of the fundamental and circulating current of MMC. It contains two cross-couplings of the d-q-axis in the fundamental current model, as well as an inter-coupling between the fundamental and circulating current models. Three controller sets have been proposed based on the detailed model. First, a detailed primary and auxiliary controller model takes into account all inter and cross-couplings. Following that, a mixed controller is obtained by combining a detailed primary and a conventional auxiliary controller. This control structure considers all coupling except primary to auxiliary inter-coupling. Exclusion of the remaining inter-coupling results in a simplified version that just considers cross-couplings. Inter-coupling from the primary to the auxiliary controller causes oscillations in the detailed controller. Inter-coupling from auxiliary to primary has no substantial advantage in a mixed controller due to its small magnitude. MMC models and controllers are simulated on RTDS for both steady-state and transient conditions. The two transient scenarios included are changes in active power and reactive power. According to the comparative performance analysis, all proposed controllers outperform the conventional controller in terms of decoupling active and reactive power. However, the detailed controller causes oscillations due to inter-coupling between the primary and auxiliary controllers. The simplified controller is free from this drawback. Apart from exhibiting improved decoupling, the proposed simplified controller also reduces the voltage stress on the SM capacitors.

Chapter 5

Analysis of Transformerless Operation of MMC

5.1 Introduction

In comparison with the conventional converters, the most significant aspect of MMC is the scalability [8]. There are several other benefits of MMC, such as higher power quality, filter-less operation, curtailed operating and maintenance expenditure, etc. [95]. Further, the MMC can operate under a transformerless configuration as well. As the cost of the transformer is significant in the converter station investment [142], the transformerless operation offers significant savings on the economic front.

The internal dynamics presented in [65], [60], and many other papers show that a third-order harmonic voltage has the potential to generate a third-order harmonic current. But most of the studies have considered either an interfacing star-delta transformer or a virtually grounded dc terminal, which restricts the flow of the third-order harmonic current. Reference [80] has developed the control of transformerless MMC-HVDC under asymmetric grid faults. A transformerless back-to-back system using MMC has been developed for interconnecting two Japanese radial feeders in [81]. These works have considered a virtually grounded dc terminal to block the third-order harmonic current. The unbalanced voltage between the two poles of the dc terminal occurs frequently [82] and continues to exist in an ungrounded MMC-HVDC system. A zero potential should be provided to ensure safe operation, which can be realised by an artificial grounding. The third-order harmonic current gets a path to flow with a physical ground connection. However, several authors have considered the interfacing transformer, which blocks the third harmonic current. In [83], a three-phase four-wire system is developed to provide a path for third harmonic currents. An analytical controller has been proposed to suppress the third harmonics. It is found that the controller relied on SM capacitor 'C_{SM}' which will be less effective due to the ageing of capacitors. Moreover, all the capacitors can not be identical too. There are several scenarios where ac and dc terminals have ground connections for providing the path to the third harmonic current.

The present chapter analyses the transformerless operation of the MMC. The updated dynamic equations for the transformerless MMC have been derived, and analyses are carried out. A closed-loop PI controller and a PR controller for suppressing the 3rd order harmonic current are developed. The performance of both controllers is compared. A state-space model has been developed and using the same the stability analysis has been carried out. The effectiveness of the proposed approach is established under steady state and transient operating conditions. The impact of the controllers on the internal dynamics of the transformerless MMC has also been investigated. A grounded dc terminal and star-connected grounded ac source are considered to create the scenarios for the analysis.

5.2 Transformerless MMC

With enough number of SMs (> 200) in each arm, synthesis of pure sinusoidal voltage waveform can be achieved, presenting the possibility of transformerless operation. The internal dynamic of MMC causes ac ripples in SM capacitor voltages, as defined in (5.1) [60, 65]. The 2nd order voltage ripple causes the 2nd order circulating current between the legs. It can be mitigated by implementing a dedicated controller scheme. Similarly, a 3rd order voltage ripple induces a 3rd order harmonic current. Under balanced operating conditions, the 3rd order currents in all three legs are in the same phase and magnitude, restricting the circulation of current between MMC legs. In addition, the delta winding of the interface transformer and the ungrounded dc terminal restrict the flow of the zero-sequence current. However, in a transformerless mode, 3rd order zero-sequence currents may flow through the ac terminal if a return path is present on the dc side.

$$u_c = \overline{u}_c + \sum_{k=1}^{\infty} u_{ck} \sin(k\omega t + \theta_k) \tag{5.1}$$

Three different scenarios have been considered to analyse the zero-sequence current and its significance in the proposed transformerless configuration. Fig. 5.1 depicts the circuits for integrating the MMC with the grid in different scenarios. The MMC's ac terminal is connected to an alternating current voltage source with a grounded neutral. A mid-point grounded dc voltage source is connected to the dc terminal. The line-to-line voltages of the MMC ac terminal and ac source are kept equal (i.e. 166 kV) to ensure that the transformer removal does not lead to voltage mismatch. To maintain consistency with the conventional setup, identical fundamental and circulating current controls have been used in each of the three systems shown in Fig. 5.1. An additional controller has not yet been implemented to handle the zero-sequence current.

5.2.1 Scenario 1

Fig. 5.1a depicts the schematic diagram of the conventional setup. The MMC is connected to the ac source through a star-delta transformer. The transformer reactance is denoted by L_t. There exists only the fundamental current i_s flowing between the ac source and the MMC terminals. Likewise, only the dc current I_{dc} exits in the dc line.

5.2.2 Scenario 2

In the second scenario shown in Fig. 5.1b, the transformer is replaced by an inductor L_{teq} with an inductance equal to the transformer inductance L_t. This provides a transformerless mode of operation while also assuring that all of the system's parametric values remain unaltered. All of the controller parameters are the same as in the preceding scenario. The elimination of the transformer enables 3rd order zero-sequence currents i_z to flow in each phase between the ac source and MMC terminals. At the midpoint of the MMC leg, it gets divided into two equal parts. One goes to the upper arm, and the other flows into the lower arm. It then continues along the dc line, where the current from all three legs combines to give $1.5i_z$ in each dc line. The 3rd-order zero-sequence harmonic current of both dc lines adds up at the grounded mid-point of the dc terminal.

5.2.3 Scenario 3

The inductor is also eliminated in the third scenario (Fig. 5.1c). Except for the inductance value, the control parameters remain the same as in the preceding scenarios. The fundamental current controller has been updated to reflect the fact that eliminating an inductor reduces the total value of inductance. Similar to the preceding case, the i_z and $1.5i_z$ currents flow in the ac and dc lines, respectively. However, omitting the inductor increases the amplitude of the current i_z as well as $1.5i_z$.

5.2.4 Analysis

A real-time digital simulator (RTDS) has been used to simulate these three scenarios. The MMCs are rated at 500 MVA and have a 166 kV line-to-line ac voltage and $\pm$160 kV dc voltage. Fig. 5.2 depicts the observation plots. Scenario 1 is represented by the solid line plot. Scenario 2 is represented by a dashed line, while Scenario 3 is represented by a dash-dot line. For the observation, the operating conditions are set to the rated ac and dc voltages, with 400 MW active power and 0 MVAr reactive

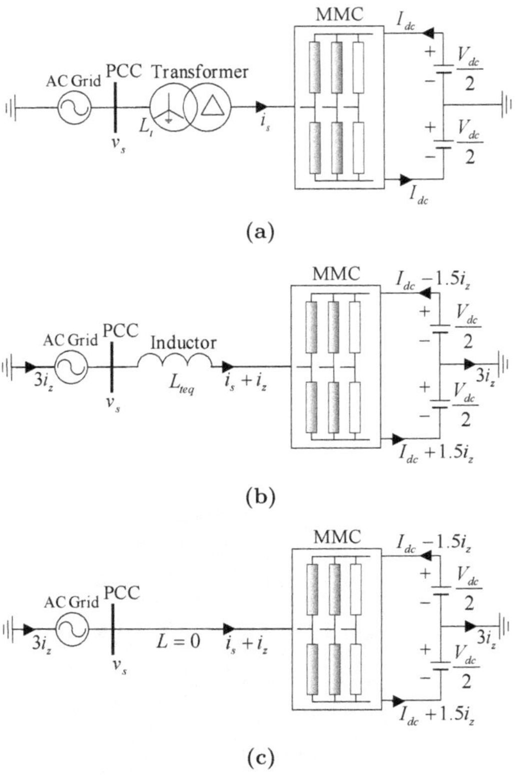

Fig. 5.1. Different circuit configurations for integrating the MMC with ac grid (a) using a star-delta transformer, (b) using a series inductor, and (C) connecting directly

power. The currents i_z and i_{zu} in Fig. 5.2 are 3rd order zero-sequence currents flowing through the phase 'a' of the ac source (i.e. ac grid) and the MMC upper arm, respectively. Since the current i_z is shared equally between the upper and lower arms, the i_{zu} equals $0.5i_z$. The current flowing over the upper dc line is a combination of dc and the sum of all three upper arm zero-sequence currents (i.e. $I_{dc} - 1.5i_z$). Likewise, the total current over the lower dc line is $I_{dc} + 1.5i_z$.

As illustrated in Fig. 5.2, in Scenario 1, with the MMC ac terminal connected to the delta side of the transformer, there is no zero-sequence current in either the ac source or the MMC arm. As a result, the dc current is obtained on the dc lines. Replacing the transformer with an equivalent inductor (Scenario 2), on the other

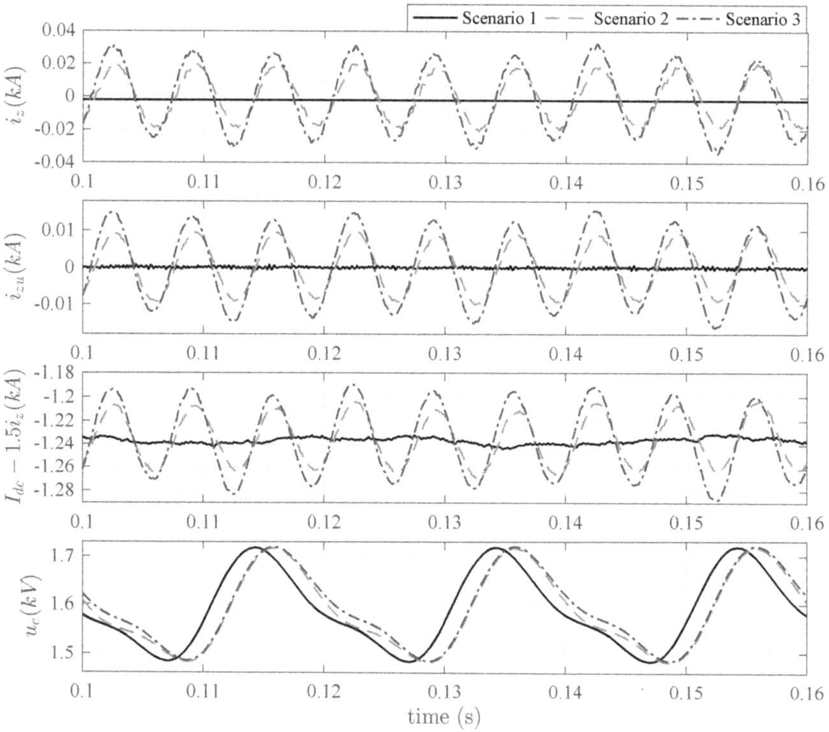

Fig. 5.2. Impact of transformerless operation of MMC variables

hand, results in a transformerless arrangement in which the zero-sequence current is not blocked. Unwanted third-order harmonic currents with amplitudes of 20 A and 10 A flow through each phase of the source and the MMC arms, respectively. On each of the dc lines, it reflects as an oscillation with 30 A amplitude. In addition to causing heat losses, this additional current leads to power oscillation on the dc line. With a dc voltage of ±160 kV, it causes a power oscillation with an amplitude of 4.8 MW on each dc line. In Scenario 3, the inductor is also eliminated, leaving MMC directly connected to the ac source. This reduces the net impedance for the 3rd order zero-sequence harmonic current while increasing the current amplitude. The i_z and i_{zu} increase to 30 A and 15 A, respectively. On the dc line, a zero-sequence current oscillation with an amplitude of 45 A occurs, resulting in a power oscillation of 7.2 MW on each dc line. The omission of the transformer results in a

30° phase difference in SM capacitor voltage. MMC phase currents exhibit a similar phase difference but are omitted from the graph for simplicity. Furthermore, the waveforms are distorted, as can be seen in Scenario 3. Thus, the removal of the transformer causes the 3rd order zero-sequence harmonic current. It flows through the source, MMC arms and the dc lines. It causes losses and power oscillation.

5.3 Internal Dynamic of MMC with 3rd Order Zero-Sequence Harmonic Current

The effect of eliminating the transformer is not limited to the flow of 3rd order zero-sequence harmonic current on the ac and dc lines. It also has an impact on the MMC's internal dynamics. The equivalent circuit diagram of the MMC for the 3rd-order harmonic current is presented in this section for study. The internal dynamics differential equations have been updated to account for the impact of the transformer's elimination.

5.3.1 Equivalent circuit for transformerless MMC

In Scenario 1, since the delta winding restricts the flow of third order harmonic current, the value of L_{teq} is infinite. In Scenario 3, however, L_{teq} equals zero. So, using Fig. 5.1b as a reference, the circuit configuration for third-order harmonic current in transformerless MMC can be depicted as in Fig. 5.3a. It has been simplified gradually in Figs. 5.3b, 5.3c, and 5.3d.

The upper and lower arms third order harmonic voltages for phase 'j' are v_{u3j} and v_{l3j}, respectively. As described in Chapter 3, these voltage components are of equal magnitude and opposing polarity. They produce the 3rd order upper and lower arm currents, i_{uzj} and i_{ljz}, respectively. Third-order currents, like voltage components, have the same magnitude and opposite polarity. The voltage components of all three phases are identical under balanced operating conditions as they are third-order components. Because they are in the same phase, the currents of all three phases

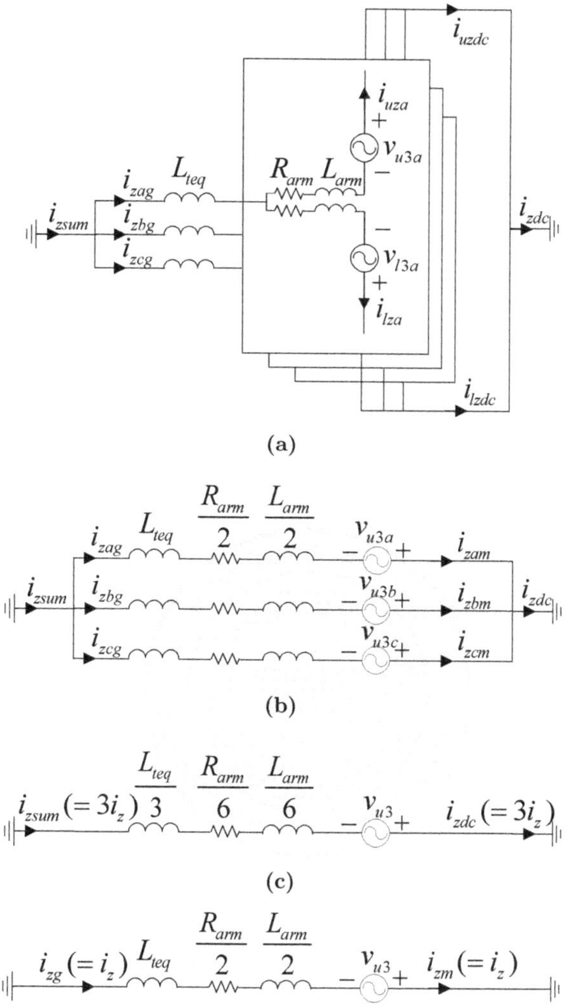

(d)

Fig. 5.3. (a)Equivalent circuit diagram of MMC for third order harmonics, (b) three-phase simplified circuit, (c) single phase equivalent.

combine to give i_{uzdc} and i_{lzdc} in the upper and lower dc lines, respectively. They are combined on the dc terminal's mid-point neutral to form i_{zdc}. i_{zjg} denotes the 3rd order currents flowing through the PCC and ac source/grid. The ac source/grid is represented by the subscript "g". i_{zsum} represents the current going through the ground on the ac terminal. The currents' mathematical relationships can be defined as in (5.2).

$$
\begin{aligned}
&i_{zsum} = \sum_{j=a,b,c} i_{zjg}; \qquad i_{uzdc} = \sum_{j=a,b,c} i_{zuj}; \qquad i_{lzdc} = \sum_{j=a,b,c} i_{zlj} \\
&i_{zjm} = i_{uzj} + i_{uzj}; \qquad i_{zdc} = i_{uzdc} + i_{lzdc}; \qquad i_{zjm} = i_{zjg} = i_{zm} = i_z;
\end{aligned}
\tag{5.2}
$$

To obtain the phase equivalent, the circuit of Fig. 5.3a is simplified into Fig. 5.3b. The parallel combinations of the arm resistance and inductance yield $R_{arm}/2$ and $L_{arm}/2$. Since the upper and lower arm voltage components are of equal magnitude, they can be represented by either of them. i_{zjm} represents the third-order current in each phase of the MMC. Since the voltage and current in all three phases are identical, their parallel combination results in Fig. 5.3c. The per phase equivalent of the same is shown in Fig. 5.3d.

5.3.2 Operation of transformerless MMC

The arm current and arm switching functions of the MMC are discussed in Chapter 3. When a third-order harmonic current is present, the upper and lower arm currents are updated as defined in (5.3). The amplitude and phase angle of the 3rd order harmonic current are represented by I_z and α_z, respectively.

$$
\begin{aligned}
i_u &= \frac{I_{dc}}{3} - \frac{I_s}{2}\sin(\omega t + \alpha_s) + I_{cir}\sin(2\omega t + \alpha_c) - \frac{I_z}{2}\sin(3\omega t + \alpha_z) \\
i_l &= \frac{I_{dc}}{3} + \frac{I_s}{2}\sin(\omega t + \alpha_s) + I_{cir}\sin(2\omega t + \alpha_c) + \frac{I_z}{2}\sin(3\omega t + \alpha_z)
\end{aligned}
\tag{5.3}
$$

To ensure efficient operation, the undesired 3rd order harmonic current components must be suppressed. To accomplish this, the arm switching functions are modified to include the 3rd order harmonic component as defined in (5.4). This component's magnitude and phase angles are denoted by V_z and β_z, respectively. The modification generates a third-order voltage with the opposite polarity as the existing third harmonic ripple voltage, thus cancelling it out. The controllers used to generate this extra function are discussed in Section 5.5.

$$
\begin{aligned}
S_u &= \frac{\frac{1}{2}V_{dc} - \frac{M}{2}V_{dc}\sin(\omega t + \beta_f) + V_{cir}\sin(2\omega t + \beta_c) - V_z\sin(3\omega t + \beta_z)}{V_{dc}} \\
S_l &= \frac{\frac{1}{2}V_{dc} + \frac{M}{2}V_{dc}\sin(\omega t + \beta_f) + V_{cir}\sin(2\omega t + \beta_c) + V_z\sin(3\omega t + \beta_z)}{V_{dc}}
\end{aligned}
\tag{5.4}
$$

5.3.3 Differential equations

The authors of [60] presented a differential equation to define the MMC's internal dynamics. The equations have been updated to include the third-order current components. The relationship between arm voltages and SM capacitor voltage, number of SMs, and arm switching function is defined in (3.1). The equation for the relationship between arm current and SM capacitor voltage is given in (3.5). Using the Park matrix P_k defined in (3.8), the fundamental and second-order components are transformed to equivalent dynamic phasors (d-q coordinate variables). The matrix P_T in (5.5) is used to derive d-q equivalents for the third-order harmonic components.

$$
P_T = \begin{bmatrix} \sin(3wt) & -\cos(3wt) \\ \cos(3wt) & \sin(3wt) \end{bmatrix} \tag{5.5}
$$

Fundamental current control governs the flow of active and reactive power. As explained in Chapter 3, the PI-based fundamental and circulating current controllers are realised in d-q coordinates. The same controllers are used here for the basic operation of the MMC.

SM capacitor voltage dynamics

Substituting (5.3) and (5.4) into (3.5) yields the comprehensive differential equation of capacitor voltage. The updated differential equations of dc, as well as different ac ripples of SM capacitor voltage, are obtained by segregating the different frequency components. These equations are transformed to d-q coordinates using (3.8), as shown in (5.6)-(5.9). The dc voltage component is denoted by $\overline{u}_c$ in this equation. u_{ckd} and u_{ckq} are the d-q components of the kth order ac voltage ripple. The fourth and higher-order components are ignored.

- dc component of SM capacitor voltage

$$\frac{d\overline{u}_c}{dt} = \frac{I_{dc}}{6C_{SM}} + \frac{I_{sd}V_{cd} + I_{sq}V_{cq}}{4C_{SM}V_{dc}} + \frac{I_{cird}V_{cird} + I_{cirq}V_{cirq}}{2C_{SM}V_{dc}} + \frac{I_{zd}V_{zd} + I_{zq}V_{zq}}{4C_{SM}V_{dc}} \tag{5.6}$$

- Fundamental frequency ripple in SM capacitor voltage

$$\begin{aligned}
\frac{du_{c1d}}{dt} &= -\omega u_{c1q} - \frac{I_{sd}}{4C_{SM}} - \frac{I_{dc}V_{cd}}{3C_{SM}V_{dc}} - \frac{I_{sd}V_{cird} + I_{sq}V_{cirq}}{4C_{SM}V_{dc}} - \frac{I_{cird}V_{cd} + I_{cirq}V_{cq}}{2C_{SM}V_{dc}} \\
&\quad - \frac{I_{zd}V_{cirq} + I_{zq}V_{cird}}{4C_{SM}V_{dc}} - \frac{I_{cird}V_{zq} + I_{cirq}V_{zd}}{2C_{SM}V_{dc}} \\
\frac{du_{c1q}}{dt} &= \omega u_{c1d} - \frac{I_{sq}}{4C_{SM}} - \frac{I_{dc}V_{cq}}{3C_{SM}V_{dc}} - \frac{I_{sd}V_{cirq} - I_{sq}V_{cird}}{4C_{SM}V_{dc}} + \frac{I_{cird}V_{cq} - I_{cirq}V_{cd}}{2C_{SM}V_{dc}} \\
&\quad - \frac{I_{zd}V_{cird} - I_{zq}V_{cirq}}{4C_{SM}V_{dc}} - \frac{I_{cird}V_{zd} - I_{cirq}V_{zq}}{2C_{SM}V_{dc}}
\end{aligned} \tag{5.7}$$

- 2nd harmonic ripple in SM capacitor voltage

$$\begin{aligned}
\frac{du_{c2d}}{dt} &= -2\omega u_{c2q} + \frac{I_{cird}}{2C_{SM}} + \frac{I_{dc}V_{cird}}{3C_{SM}V_{dc}} + \frac{I_{sd}V_{cd} - I_{sq}V_{cq}}{4C_{SM}V_{dc}} + \frac{I_{sd}V_{zq} + I_{sq}V_{zd}}{4C_{SM}V_{dc}} \\
&\quad + \frac{I_{zd}V_{cq} + I_{zq}V_{cd}}{4C_{SM}V_{dc}} \\
\frac{du_{c2q}}{dt} &= 2\omega u_{c2d} + \frac{I_{cirq}}{2C_{SM}} + \frac{I_{dc}V_{cirq}}{3C_{SM}V_{dc}} + \frac{I_{sd}V_{cq} + I_{sq}V_{cd}}{4C_{SM}V_{dc}} + \frac{I_{sd}V_{zd} - I_{sq}V_{zq}}{4C_{SM}V_{dc}} \\
&\quad + \frac{I_{zd}V_{cd} - I_{zq}V_{cq}}{4C_{SM}V_{dc}}
\end{aligned} \tag{5.8}$$

- 3rd harmonic ripple in SM capacitor voltage

$$\begin{aligned}
\frac{du_{c3d}}{dt} &= 3\omega u_{c3q} - \frac{I_{zd}}{4C_{SM}} - \frac{I_{dc}V_{zd}}{3C_{SM}V_{dc}} - \frac{I_{sd}V_{cirq} + I_{sq}V_{cird}}{4C_{SM}V_{dc}} - \frac{I_{cird}V_{cq} + I_{cirq}V_{cd}}{2C_{SM}V_{dc}} \\
\frac{du_{c3q}}{dt} &= -3\omega u_{c3d} - \frac{I_{zq}}{4C_{SM}} - \frac{I_{dc}V_{zq}}{3C_{SM}V_{dc}} - \frac{I_{sd}V_{cird} - I_{sq}V_{cirq}}{4C_{SM}V_{dc}} - \frac{I_{cird}V_{cd} - I_{cirq}V_{cq}}{2C_{SM}V_{dc}}
\end{aligned} \tag{5.9}$$

Arm voltage

The updated expression for arm voltage is obtained by substituting (5.4) and (5.1) into (3.1). The segregation of different frequency components results in the respective components of arm voltage (5.10). v_{udc} and v_{uk} are dc and k^{th}-order harmonic components of the arm voltages, respectively. Lower arm voltage can also be expressed in a similar way. The polarity of the dc and even order components of the upper and lower arm voltages is the same, whereas the polarity of the odd-order components of both arms is opposite. The fourth and higher-order components are ignored.

$$
\begin{aligned}
v_{udc} &= \frac{N\overline{u}_c}{2} - \frac{MNu_{c1}}{4}\cos(\beta_f - \theta_1) + \frac{Nu_{c2}V_{cir}}{2V_{dc}}\cos(\beta_c - \theta_2) - \frac{Nu_{c3}V_z}{2V_{dc}}\cos(\beta_z - \theta_3) \\
v_{u1} &= \frac{Nu_{c1}}{2}\sin(\omega t + \theta_1) - \frac{MNu_{c2}}{4}\cos(\omega t - \beta_f + \theta_2) - \frac{MN\overline{u}_c}{2}\sin(\omega t + \beta_f) \\
&+ \frac{Nu_{c1}V_{cir}}{2V_{dc}}\cos(\omega t + \beta_c - \theta_1) + \frac{Nu_{c3}V_{cir}}{2V_{dc}}\cos(\omega t - \beta_c + \theta_3) \\
&- \frac{Nu_{c2}V_z}{2V_{dc}}\cos(\omega t + \beta_z - \theta_2) \\
v_{u2} &= \frac{MNu_{c1}}{4}\cos(2\omega t + \beta_f + \theta_1) + \frac{Nu_{c2}}{2}\sin(2\omega t + \theta_2) - \frac{MNu_{c3}}{4}\cos(2\omega t - \beta_f + \theta_3) \\
&+ \frac{N\overline{u}_cV_{cir}}{V_{dc}}\sin(2\omega t + \beta_c) - \frac{Nu_{c1}V_z}{2V_{dc}}\cos(2\omega t + \beta_z - \theta_1) \\
v_{u3} &= \frac{MNu_{c2}}{4}\cos(3\omega t + \beta_f + \theta_2) + \frac{Nu_{c3}}{2}\sin(3\omega t + \theta_3) - \frac{Nu_{c1}V_{cir}}{2V_{dc}}\cos(3\omega t + \beta_c + \theta_1) \\
&- \frac{N\overline{u}_cV_z}{V_{dc}}\sin(3\omega t + \beta_z)
\end{aligned}
\tag{5.10}
$$

Current dynamics

By applying KVL for different frequency components, the equations for dc, fundamental and 2nd order current are derived, respectively, as in (3.10), (3.11), and (3.12). Similarly, as we apply the KVL to the third order component in Fig. 5.3d, we get (5.11).

$$\frac{di_z}{dt} = \frac{1}{L_{eq}} v_{u3} - \frac{R_{eq}}{L_{eq}} i_z \tag{5.11}$$

Substituting the value of the dc component of upper arm voltage v_{udc} form (5.10) in (3.10) yields the updated dc current dynamics (5.12).

$$\begin{aligned}\frac{dI_{dc}}{dt} = &-\frac{I_{dc}R_{arm}}{L_{arm}} + \frac{3(V_{dc} - N\overline{u}_c)}{2L_{arm}} + \frac{3N(u_{c1d}V_{cd} + u_{c1q}V_{cq})}{2L_{arm}V_{dc}} \\ &- \frac{3N(u_{c2d}V_{cird} + u_{c2q}V_{cirq})}{2L_{arm}V_{dc}} + \frac{3N(u_{c3d}V_{zd} + u_{c3q}V_{zq})}{2L_{arm}V_{dc}}\end{aligned} \tag{5.12}$$

Similarly, replacing the ac voltage components v_{u1}, v_{u2}, and v_{u2} from (5.10) to (3.11), (3.12), and (5.11) yields the updated dynamic of the ac current components. Using the P_k matrix for fundamental and 2nd order circulating components and the P_T matrix for 3rd order zero-sequence, the differential equations of the respective component in d-q coordinates are derived as presented in (5.13)-(5.15).

$$\begin{aligned}\frac{dI_{sd}}{dt} &= -\omega I_{sq} + \frac{V_{sd} - I_{sd}R_{eq}}{L_{eq}} + \frac{Nu_{c1d}}{2L_{eq}} - \frac{N\overline{u}_cV_{cd}}{L_{eq}V_{dc}} + \frac{N(u_{c1d}V_{cird} + u_{c1q}V_{cirq})}{2L_{eq}V_{dc}} \\ &+ \frac{N(u_{c3d}V_{cirq} + u_{c3q}V_{cird})}{2L_{eq}V_{dc}} - \frac{N(u_{c2d}V_{cd} + u_{c2q}V_{cq} + u_{c2d}V_{zq} + u_{c2q}V_{zd})}{2L_{eq}V_{dc}} \\ \frac{dI_{sq}}{dt} &= \omega I_{sd} + \frac{V_{sq} - I_{sq}R_{eq}}{L_{eq}} + \frac{Nu_{c1q}}{2L_{eq}} - \frac{N\overline{u}_cV_{cq}}{L_{eq}V_{dc}} + \frac{N(u_{c1d}V_{cirq} - u_{c1q}V_{cird})}{2L_{eq}V_{dc}} \\ &+ \frac{N(u_{c3d}V_{cird} - u_{c3q}V_{cirq})}{2L_{eq}V_{dc}} + \frac{N(u_{c2d}V_{cq} - u_{c2q}V_{cd} - u_{c2d}V_{zd} + u_{c2q}V_{zq})}{2L_{eq}V_{dc}}\end{aligned} \tag{5.13}$$

$$\begin{aligned}\frac{dI_{cird}}{dt} &= -2\omega I_{cirq} - \frac{I_{cird}R_{arm}}{L_{arm}} - \frac{Nu_{c2d}}{2L_{arm}} - \frac{N\overline{u}_cV_{cird}}{L_{arm}V_{dc}} + \frac{N(u_{c1d}V_{cd} - u_{c1q}V_{cq})}{2L_{arm}V_{dc}} \\ &+ \frac{N(u_{c3d}V_{cq} + u_{c3q}V_{cd})}{2L_{arm}V_{dc}} + \frac{N(u_{c1d}V_{zq} + u_{c1q}V_{zd})}{2L_{arm}V_{dc}} \\ \frac{dI_{cirq}}{dt} &= 2\omega I_{cird} - \frac{I_{cirq}R_{arm}}{L_{arm}} - \frac{Nu_{c2q}}{2L_{arm}} - \frac{N\overline{u}_cV_{cirq}}{L_{arm}V_{dc}} + \frac{N(u_{c1d}V_{cq} + u_{c1q}V_{cd})}{2L_{arm}V_{dc}} \\ &+ \frac{N(u_{c3d}V_{cd} - u_{c3q}V_{cq})}{2L_{arm}V_{dc}} + \frac{N(u_{c1d}V_{zd} - u_{c1q}V_{zq})}{2L_{arm}V_{dc}}\end{aligned} \tag{5.14}$$

$$\frac{dI_{zd}}{dt} = 3\omega I_{zq} - \frac{I_{zd}R_{eq}}{L_{eq}} + \frac{Nu_{c3d}}{2L_{eq}} - \frac{N\overline{u}_c V_{zd}}{L_{eq}V_{dc}} + \frac{N(u_{c1d}V_{cirq} + u_{c1q}V_{cird})}{2L_{eq}V_{dc}} - \frac{N(u_{c2d}V_{cq} + u_{c2q}V_{cd})}{2L_{eq}V_{dc}}$$
$$\frac{dI_{zq}}{dt} = -3\omega I_{zd} - \frac{I_{zq}R_{eq}}{L_{eq}} + \frac{Nu_{c3q}}{2L_{eq}} - \frac{N\overline{u}_c V_{zq}}{L_{eq}V_{dc}} + \frac{N(u_{c1d}V_{cird} - u_{c1q}V_{cirq})}{2L_{eq}V_{dc}} - \frac{N(u_{c2d}V_{cd} - u_{c2q}V_{cq})}{2L_{eq}V_{dc}} \tag{5.15}$$

5.4 Impact on Power Transfer Capability

The arm energy and power transfer capability of the MMC are functions of arm voltage and arm current [139]. In transformerless operation, the arm current has an additional component of the 3rd order harmonic current. Therefore, its impact on the arm energy and power transfer capability is inevitable.

5.4.1 Required energy per arm

In the present study, the ac source voltage and current are defined as follows:

$$v_{sj} = V_s \sin(\omega t); \qquad i_{sj} = I_s \sin(\omega t + \alpha_s) \tag{5.16}$$

The V_s and I_s are the amplitude of the ac source voltage and current, respectively. The $_s$ is phase angle of the fundamental current. The upper arm voltage including voltage across arm resistance (R_{arm}) and arm inductance (L_{arm}) is defined as:

$$v_{uarm} = \frac{1}{2}V_{dc}\big[1 - m\sin(\omega t)\big] \tag{5.17}$$

Equation (5.3) defines the upper arm current. The dc current component can be defined in terms of fundamental component as in (5.18) [143]. Likewise, the magnitude of the 2nd and 3rd order current components can be defined as a function of the magnitude of the fundamental component as defined in (5.19). Here, K_{cir}, and K_z are proportionality constants. Substituting (5.18) and (5.19) in (5.3), yields

(5.20). Since, a dedicated 2nd order controller is in place to eliminate the circulating current, the K_{cir} becomes almost zero.

$$\frac{I_{dc}}{3} = \frac{-mI_s}{4}\cos\alpha_s; \tag{5.18}$$

$$I_{cir} = \frac{K_{cir}I_s}{2} \qquad I_z = K_z I_s \tag{5.19}$$

$$i_u = -\frac{I_s}{2}\left[\frac{m}{2}\cos\alpha_s + \sin(\omega t + \alpha_s) - K_{cir}\sin(2\omega t + \alpha_c) + K_z\sin(3\omega t + \alpha_z)\right] \tag{5.20}$$

The instantaneous power is derived from upper arm voltage and current as in (5.21). Expanding the same yields (5.22).

$$\begin{aligned} P_{uz}(t) &= v_{uarm} * i_u \\ P_{uz}(t) &= -\frac{1}{2}V_{dc} * \frac{I_s}{2}m * \left[1 - m\sin(\omega t)\right] * \left[\frac{m}{2}\cos\alpha_s + \sin(\omega t + \alpha_s)\right. \\ &\quad \left. - K_{cir}\sin(2\omega t + \alpha_c) + K_z\sin(3\omega t + \alpha_z)\right] \end{aligned} \tag{5.21}$$

$$\begin{aligned} P_{uz}(t) = -\frac{1}{3}P_s\Bigg[& \frac{1}{m}\sin(\omega t + \alpha_s) - \frac{m}{2}\cos(\alpha_s)\sin(\omega t) + \frac{K_{cir}}{2}\cos(\omega t + \alpha_c) \\ & - \frac{K_{cir}}{m}\sin(2\omega t + \alpha_c) + \frac{1}{2}\cos(2\omega t + \alpha_s) - \frac{K_z}{2}\cos(2\omega t + \alpha_z) \\ & + \frac{K_z}{m}\sin(3\omega t + \alpha_z) - \frac{K_{cir}}{2}\cos(3\omega t + \alpha_c) + \frac{K_z}{2}\cos(4\omega t + \alpha_z)\Bigg] \end{aligned} \tag{5.22}$$

where $P_s = \frac{3}{2}v_s i_s$

The integration of the upper arm instantaneous power yields the energy requirement of the upper arms as defined in (5.23). Similarly, the instantaneous power and energy requirement of the lower arm also can be derived.

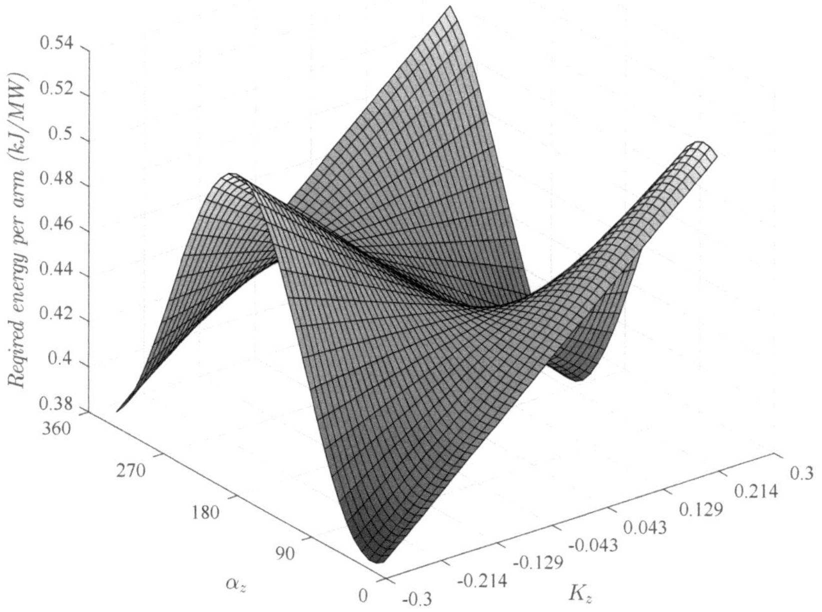

Fig. 5.4. Variation in energy requirement per arm with respect to third-order zero-sequence current harmonic

$$\begin{aligned} e_{uz}(t) =& \frac{1}{3\omega} P_s \Bigg[\frac{1}{m} \cos(\omega t + \alpha_s) - \frac{m}{2} \cos(\alpha_s) \cos(\omega t) - \frac{K_{cir}}{2} \sin(\omega t + \alpha_c) \\ & - \frac{K_{cir}}{2m} \cos(2\omega t + \alpha_c) - \frac{1}{4} \sin(2\omega t + \alpha_s) + \frac{K_z}{4} \sin(2\omega t + \alpha_z) \\ & + \frac{K_z}{3m} \cos(3\omega t + \alpha_z) + \frac{K_{cir}}{6} \sin(3\omega t + \alpha_c) - \frac{K_z}{8} \sin(4\omega t + \alpha_z) \Bigg] \end{aligned} \tag{5.23}$$

The relation of the arm energy requirement with 3rd order harmonic current is presented in Fig. 5.4. The modulation index is set to 0.8. The proportionality constant (K_z) is varied from -0.3 to 0.3 and the phase angle α_z is varied from 0° to

360°, with respect to the phase angle of fundamental current (α_s). For α_z near 0° and 360°, with an increase in K_z, the required energy per arm also increases. However, for α_z near 180°, the effect is reversed. The increase in arm energy requires an increase in the size of the capacitor to handle the additional energy. Thus, the transformerless operation increases the arm energy requirement of the MMC.

5.4.2 Operating range

The power transfer capability of the MMC can be obtained from the following equations:

$$P_{capability}(\alpha_s) \leq \frac{\Delta E_{max}}{max(|e_{uz}(t)|)} \times P_s \cos(\alpha_s) \tag{5.24}$$

$$Q_{capability}(\alpha_s) \leq \frac{\Delta E_{max}}{max(|e_{uz}(t)|)} \times P_s \sin(\alpha_s) \tag{5.25}$$

where ΔE_{max} is the maximum excess arm energy capability [139], which can be defined as in (5.26).

$$\Delta E_{max} = (k_{max}^2 - 1)\frac{N}{2} C_{SM} * u_{c,avg}^2 \tag{5.26}$$

Fig. 5.5 depicts the power transfer capability of transformerless operation. To eliminate the impact of 2nd order circulating current, the K_{cir} is set to zero. The rated active power of the MMC is 500 MW, as shown by the solid vertical line while the dotted line represents the apparent power. The effective operating range of the system under transformerless mode with $K_z = 0$ is depicted by a dash-dot line. However, with the presence of the third-order zero-sequence harmonic current, the range is adversely affected. The plot for the positive value of K_z is presented by the solid line. The dashed lines depict the values of the negative K_z value. The plots are included for the I_z equal to $\pm 10\%$, $\pm 20\%$ and $\pm 30\%$ of fundamental current I_s.

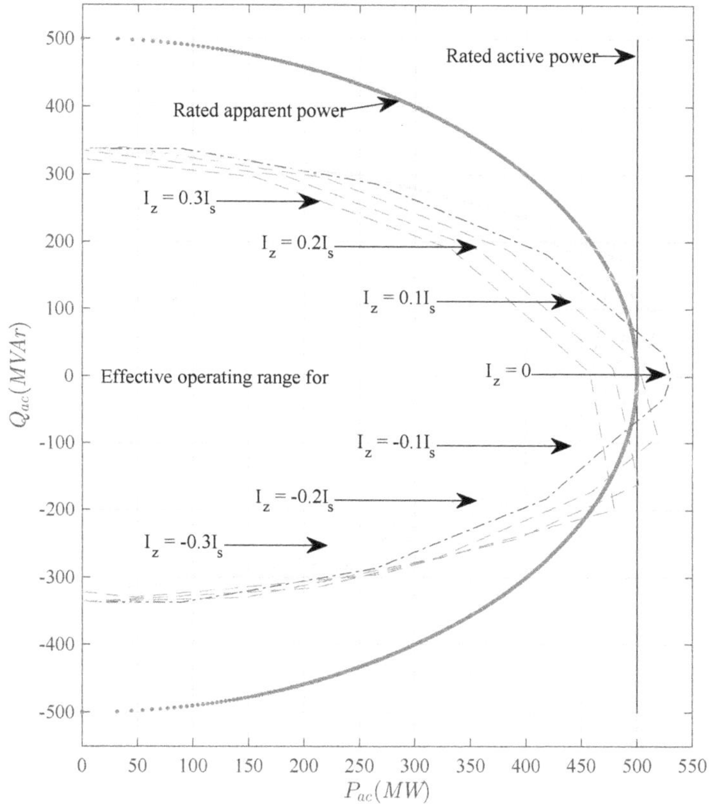

Fig. 5.5. Power transfer capability for transformerless operation.

5.5 Third-Order Harmonic Current Suppression

The presence of the 3rd order zero-sequence current in transformerless MMC results in power loss as well as power oscillation over the dc lines. To ensure the safe operation under the proposed transformerless mode, suppression of the unwanted zero-sequence current is required. Scenario 3 of Fig. 5.1 has been considered for the controller design, as it has no interfacing element which leads to the most severe i_z among all three configurations. Fig. 5.3d presents the equivalent circuit diagram for 3rd order harmonic current. The two controllers designed for suppressing i_z are closed-loop PR and PI controllers. Similar to primary and secondary controllers for fundamental and circulating current, the proposed controller can be termed a tertiary controller.

5.5.1 PR controller

Fig. 5.6 depicts the proposed proportional resonant (PR) controller. The transfer function of the closed-loop PR controller is defined as (5.27). The controller parameters are listed in Table 5.1.

$$G_{PR}(s) = K_P + \frac{K_r s}{s^2 + \omega_o^2} \tag{5.27}$$

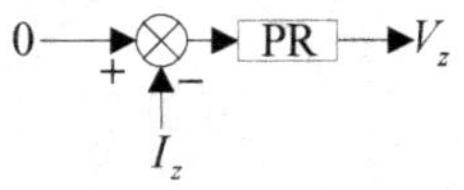

Fig. 5.6. Proposed PR controller for suppressing third-order zero-sequence harmonic current

The bode plot for the PR controller is shown by the solid line in Fig. 5.8. The magnitude plot shows a spike at ω_o. Likewise, the phase plot observes phase-shift. The step response of the PR controller is faster than the step response of the PI controller.

Table 5.1: System and tertiary controller parameters

System parameter		PI controller parameters	
Rated power	500 MVA	K_{pz}	30
ac voltage (line-to-line)	166 kV	K_{iz}	30
dc voltage (pole-to-pole)	320 kV	PR controller parameters	
Number of SMs per arm	200	K_p	200
Arm resistance	1 Ω	K_r	3000
Arm inductance	60 milihenry	ω_o	942
SM capacitance	1000 microfarad		

5.5.2 PI controller

Fig. 5.7 presents the proposed d-q coordinate-based PI controller. The references are set to zero, and the input signals, I_{zd} and I_{zq} are deduced using the P_T matrix from (5.5). The system and tertiary PI controller parameters are presented in Table 5.1. Further, the bode plot and the step response of the system with the PI controller are depicted in Fig. 5.8 and Fig. 5.9, respectively. Both the plots suggest a stable operation with the PI controller.

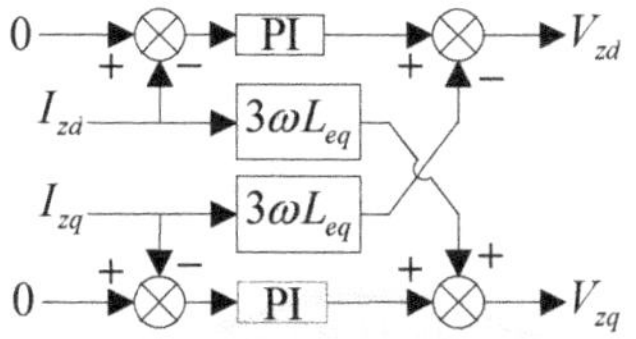

Fig. 5.7. Proposed PI controller for suppressing third-order zero-sequence harmonic current

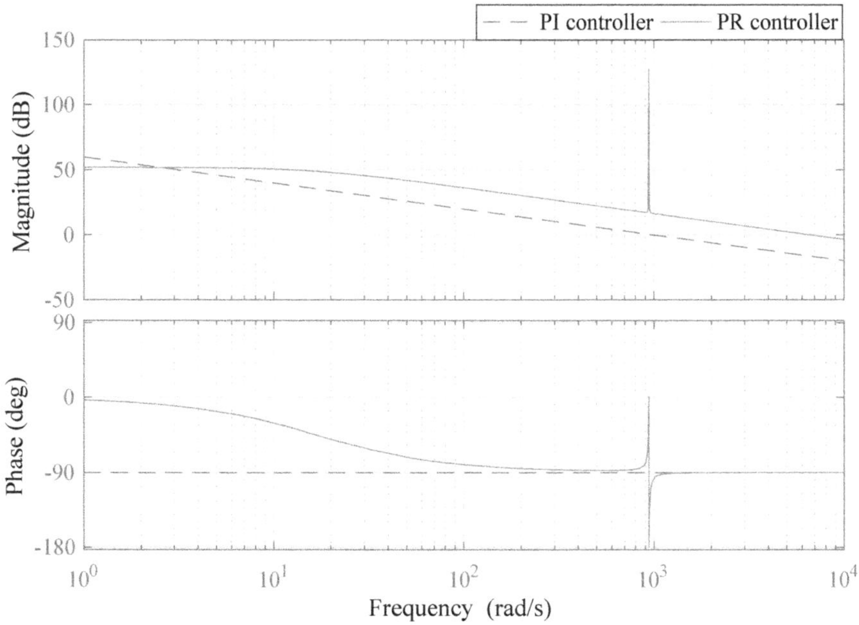

Fig. 5.8. Bode plot with PI and PR controller

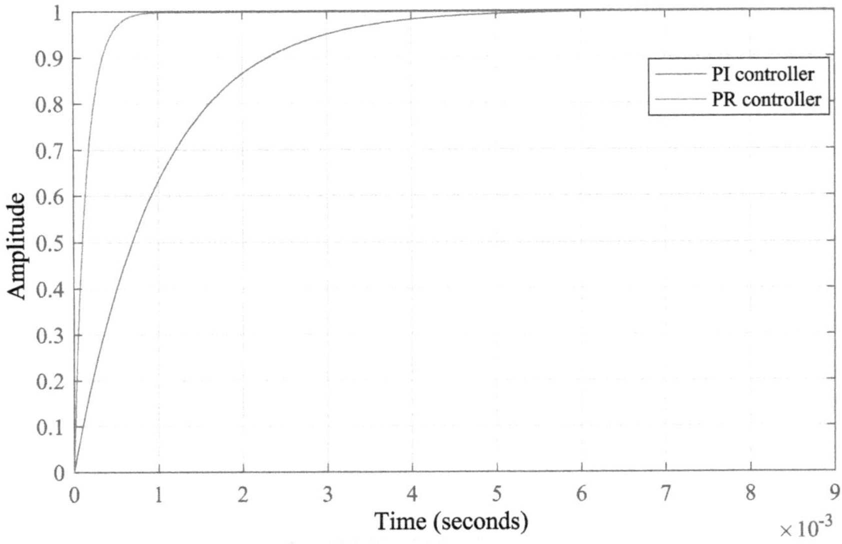

Fig. 5.9. Step response with PI and PR controller

Adding the control outputs of the tertiary controllers (PI or PR) to the conventional arm switching function, the updated arm switching function, defined in (5.4), is obtained. The added third harmonic term suppresses the undesired current component by modifying the arm voltage of the MMC. Since the different components are intertwined through the SM capacitor voltage, the i_z affects the complete internal dynamics as presented in the previous section. To implement the different controllers, the current measurements are derived from the measured value of upper and lower arm currents as defined in (5.28). Fig. 5.10 presents the pictorial depiction of the same.

$$\begin{aligned} I_s \sin(wt + \alpha_s) &= i_l - i_u - I_z \sin(3wt + \alpha_z) \\ I_z \sin(3wt + \alpha_z) &= \frac{1}{3} \sum_{j=a,b,c} i_{lj} - i_{uj} \\ I_{cir} \sin(2wt + \alpha_c) &= \frac{i_u + i_l}{2} - \frac{I_{dc}}{3} \\ I_{dc} &= \frac{1}{3} \sum_{j=a,b,c} \frac{i_{uj} + i_{lj}}{2} \end{aligned} \tag{5.28}$$

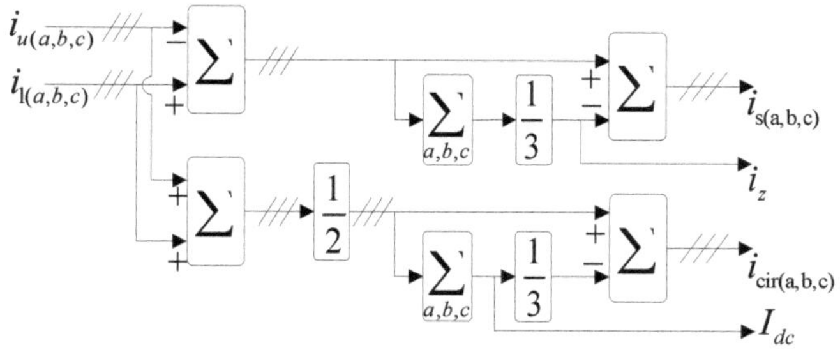

Fig. 5.10. Deducing different current components from arm currents

5.6 Stability Analysis

To analyse the stability, a state-space model has been developed. Under transformerless configuration, MMC gives a 14th order state space model. Adding the ac source, primary and secondary controllers, a 20th order system is obtained. Implementation of tertiary PI control adds two more poles. Linearisation yields the state-space model as shown in (5.29).

$$\mathbf{\Delta \dot{X} = A\Delta x + B\Delta u} \tag{5.29}$$

$$\mathbf{\Delta x} = [\mathbf{\Delta \overline{u}_c}, \mathbf{\Delta u_{c1d}}, \mathbf{\Delta u_{c1q}}, \mathbf{\Delta u_{c2d}}, \mathbf{\Delta u_{c2q}}, \mathbf{\Delta u_{c3d}}, \mathbf{\Delta u_{c3q}}, \mathbf{\Delta I_{dc}}, \mathbf{\Delta I_{sd}}, \mathbf{\Delta I_{sq}}, \mathbf{\Delta I_{cird}},$$

$$\mathbf{\Delta I_{cirq}}, \mathbf{\Delta I_{zd}}, \mathbf{\Delta I_{zq}}, \mathbf{\Delta y_{sd}}, \mathbf{\Delta y_{sq}}, \mathbf{\Delta y_{cd}}, \mathbf{\Delta y_{cq}}, \mathbf{\Delta y_{zd}}, \mathbf{\Delta y_{zq}}, \mathbf{\Delta y_{p}}, \mathbf{\Delta \theta_{PLL}}]^{\mathbf{T}}$$

$$\mathbf{\Delta u} = [\mathbf{\Delta P^*_{ac}}, \mathbf{\Delta Q^*_{ac}}, \mathbf{\Delta V_{dc}}, \mathbf{\Delta \omega}]^T$$

where $\mathbf{A}$ and $\mathbf{B}$ represent the state matrix and input matrix, respectively. $\mathbf{\Delta x}$ is the state vector and $\mathbf{\Delta u}$ is the input vector. State variables $\mathbf{\Delta y_{zd}}$, and $\mathbf{\Delta y_{zq}}$ are from the tertiary PI controller. Other variables are the same as defined in Chapter 3.

Fig. 5.11 and Fig. 5.12 present the comparative plots of EMT and space-space model. The solid line represents the EMT model developed in RTDS, and the state-space model in MATLAB is depicted by the dashed line. Fig. 5.11 depicts the plots for transformerless MMC without a tertiary controller. Likewise, the plots for the case with the tertiary PI controller are presented in Fig. 5.12. In both cases, initially, the MMCs are operating at 400 MW and 0 MVAr with all voltages at their rated value. At a time instant of 1.8 seconds, the active power has been increased by 20 MW. The EMT as well as state-space plots of upper arm current i_u and SM capacitor voltage u_c trace the same path. The circulating current increases at the instant of change for both models and then are eliminated by the secondary controller. The plots of transformerless MMC without controller shows the presence of 3rd order zero-sequence current (i_z) as shown in Fig. 5.11. The same is eliminated when a tertiary PI controller is applied, as presented in Fig. 5.12. As shown in the plots, EMT, as well as the state-space model, are in agreement. Thus, the developed model is validated.

The Eigenvalues for the transformerless MMC without a tertiary controller and with a tertiary PI controller are presented in Table 5.2. All the Eigenvalues are in

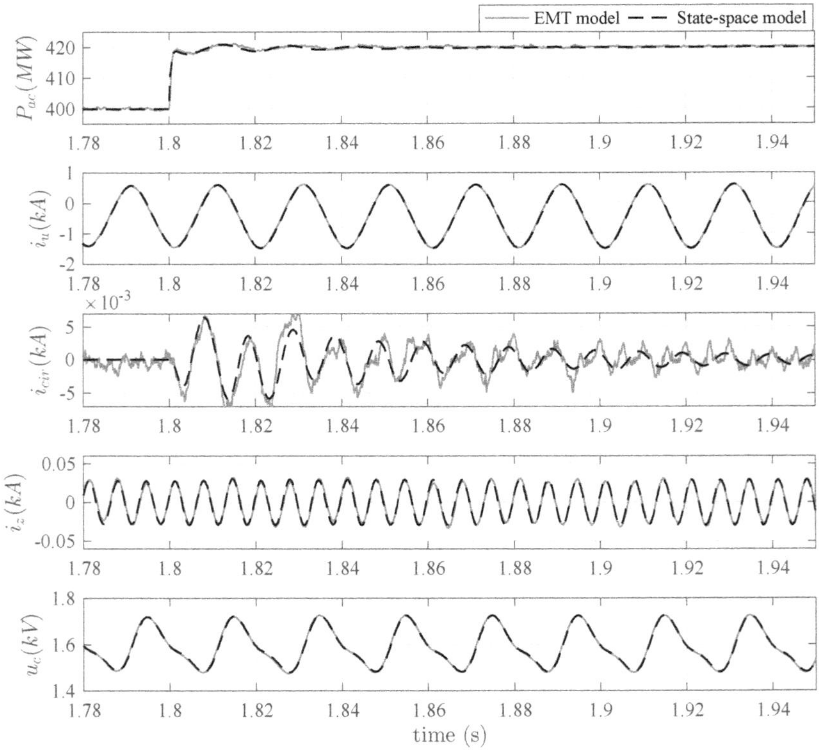

Fig. 5.11. EMT and small-signal analysis without tertiary controller

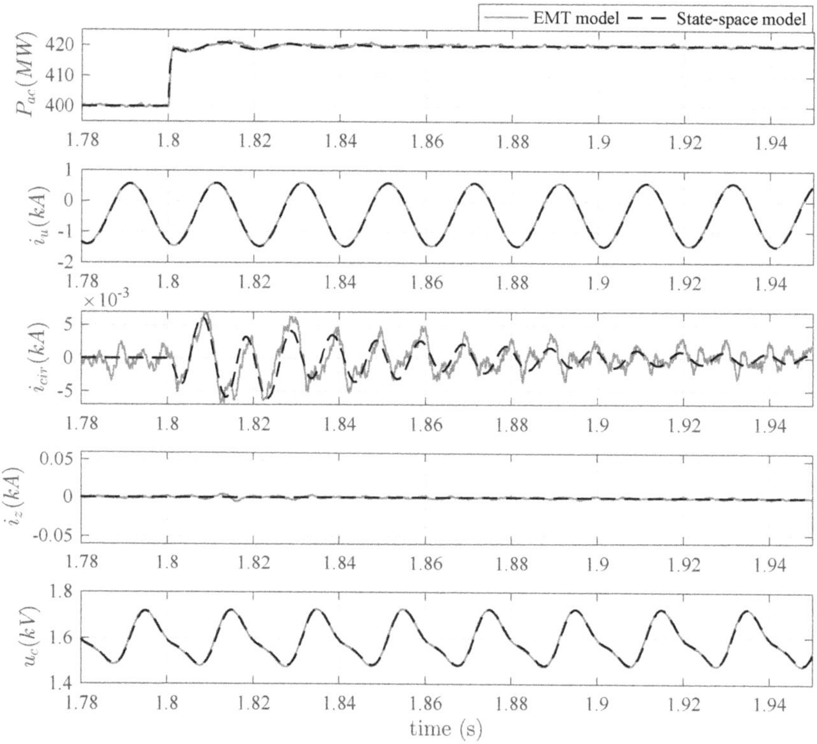

Fig. 5.12. EMT and small-signal analysis with tertiary PI controller

Table 5.2: Eigenvalues for transformerless operation

	Without tertiary controller			With tertiary PI controller		
Modes	Eigenvalues	f(Hz)	ζ	Eigenvalues	f(Hz)	ζ
1	-27.97±3599	572	0.008	-61.76±2876	458	0.021
2	-28.26±2108	335	0.013	-8229±1445	230	0.98
3	-913.5±74.8	11.91	0.997	-921±76.12	12.12	0.997
4	-78.61±637	101.4	0.122	-76.16±672	107	0.113
5	-37.40±413	65.70	0.090	-38.37±412	65.58	0.093
6	-17.17±230	36.54	0.077	-17.92±229	36.51	0.078
7	-15.71±3.51	0.56	0.976	-15.69±3.19	0.51	0.980
8	-6.65±0.52	0.08	0.997	-6.65±0.52	0.08	0.997
9	-1355374	0	1	-1355374	0	1
10	-10	0	1	-10	0	1
11	-2463	0	1	-2463	0	1
12	-2467	0	1	-2467	0	1
13				-8.71±1.83	0.29	0.979

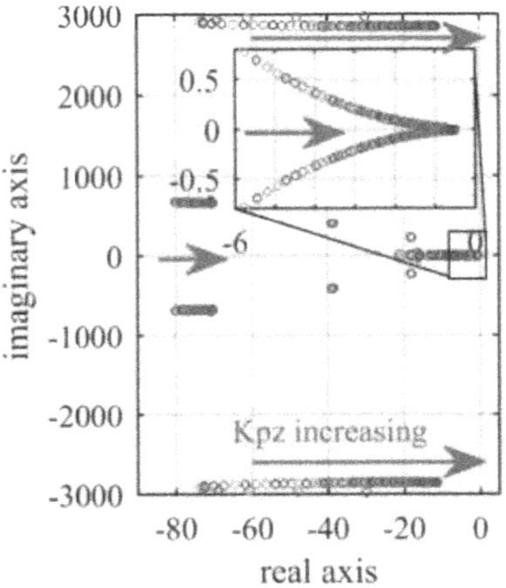

(a)

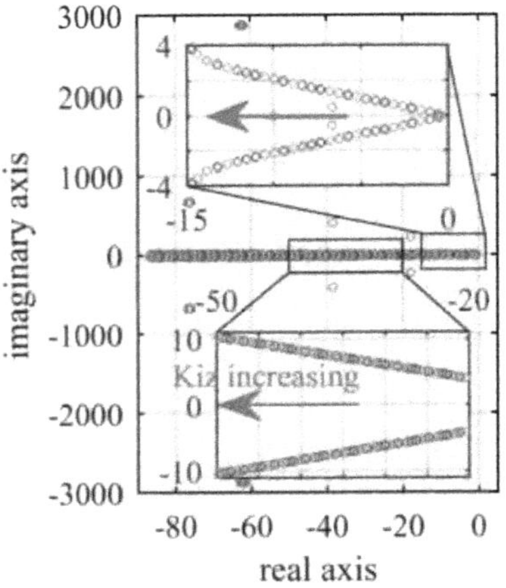

(b)

Fig. 5.13. Root-locus for tertiary PI controller variables (a) $5 < K_{pz} < 2000$ **(b)** $30 < K_{iz} < 5000$

the negative plane and both MMCs are stable. The damping ratio (ζ) of the modes has significantly increased with the implementation of the proposed controller. The damping ratio of mode 2 has been significantly increased. The tertiary controller adds an additional mode, with very high damping (i.e. 0.979). Thus, the PI controller enhances the stability of the system. Fig. 5.13 presents the root-locus plot for the parameter variations in the tertiary PI controller. The proportional gain (K_{pz}) is varied from 5 to 2000 and the pole movement is recorded in Fig. 5.13a. With an increase in the K_{pz}, the poles move towards zero, however, they do not cross to the positive plane. In Fig. 5.13b, pole movements for variation of integral gain (K_{iz}) from 30 to 5000 have been shown. With an increase in K_{iz}, the poles move away from zero. Thus, the proposed PI controller gives a stable operation over a wide range.

5.7 Results and Discussions

The performances of the closed-loop tertiary PI and PR controllers for suppressing the unwanted 3rd order zero-sequence current have been investigated. The EMT models of three transformerless MMCs (without a controller, with a PR controller and with a PI controller) have been implemented on the real-time digital simulator (RTDS). The parameters for the fundamental and circulating current controller are not changed and are the same as for the conventional MMC configuration in Scenario 1 (Fig. 5.1c). Initial operating conditions are set to 400 MW active power and 0 MVAr reactive power, with rated voltages.

Figs. 5.14-5.16 give the performance analysis of both controllers. The solid lines represent the MMC without a controller; the dashed line depicts the PR controller and the dash-dot line is for the PI controller. Fig. 5.14 presents the performance of controllers under steady-state operation whereas Figs.5.15 and 5.16 show the transient operating conditions. As presented in Section 5.3, due to the intertwined nature of variables, the zero-sequence current and voltage affect all the other variables too. Thus, the impact analysis of the proposed controller on primary and secondary controllers becomes very important. Fig. 5.17 presents the individual harmonic distortion (IHD) of the 3rd order harmonic current. Further, to assert that the tertiary controllers do not affect the internal dynamics adversely, the arm current and SM capacitor voltage have been analysed in Fig. 5.18.

5.7.1 Steady state operation

Under the steady-state operation as shown in Fig. 5.14, the active power (P_{ac}) plots and the reactive power (Q_{ac}) plots for all three control modes overlap with each other. It ensures that all three MMCs with their respective controllers are functioning correctly. The I_{sd} and I_{sq} represent the fundamental current in d-q coordinates. The presence of a third-order harmonic current does not have any effect on the fundamental current of the MMC. Similar to power plots, current plots for all three control modes overlap with each other. The zero value of the circulating current i_{cir} signifies the suppression of the circulating current in all three models. Plots of i_z and P_z represent the 3rd order zero-sequence harmonic phase current flowing through the source and the 3rd order power oscillation over the dc line, respectively. P_z has been calculated using (5.30).

$$P_z = V_{dc} * (I_{dc} - 1.5i_z) - V_{dc} * I_{dc} \tag{5.30}$$

When the controller is not used, the current i_z, half of which passes through each arm of the MMC leg, reaches 25-30 A, leading to the power oscillation of 12-14.4 MW, on the dc terminal. Applying the tertiary controller (PR/PI) successfully eliminates this unwanted current, as well as dc line power oscillation. Both controllers are equally efficient in achieving the objective.

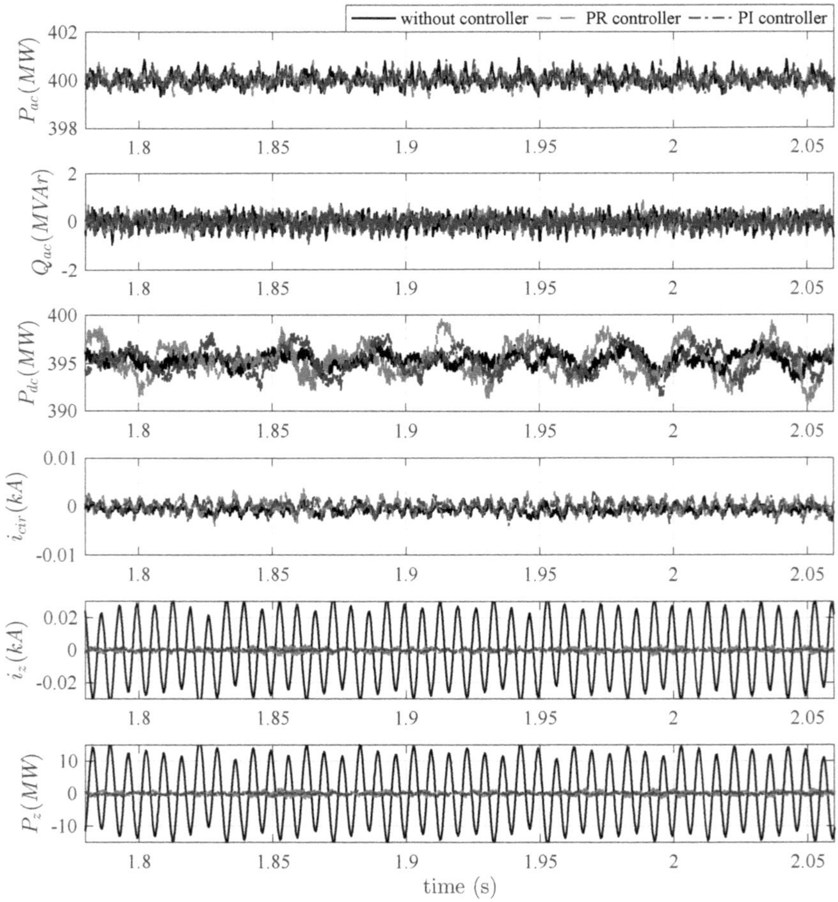

Fig. 5.14. Controller performance analysis under steady-state operation.

5.7.2 Transient operation: 20 MW change in active power

As depicted in Fig. 5.15, initially the MMCs are operating at the rated voltage with 400 MW active power and 0 MVAr reactive power. At time instant 1.8 seconds, the active power reference has been increased by 20 MW. All three MMCs give the overlapping waveforms for P_{ac}. However, the plots for MMCs with the tertiary controller, have a slightly higher peak at the instant of change. Due to the coupling effect, change in active power causes oscillations in reactive power too, which are identical and die out gradually. Application of the tertiary controller leads to a

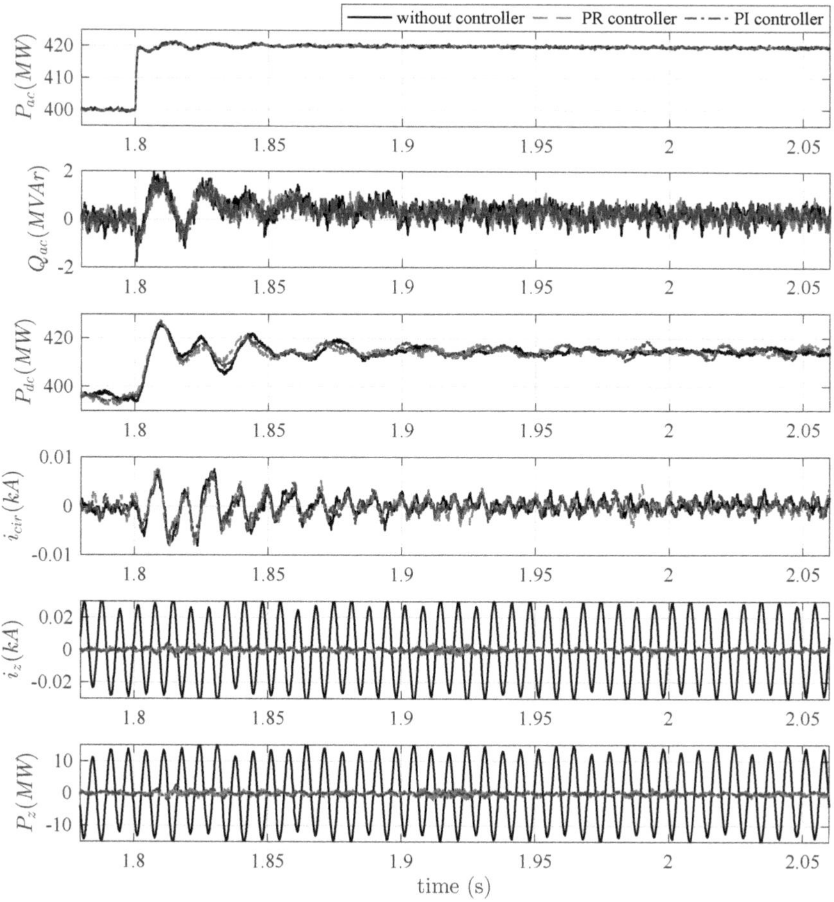

Fig. 5.15. Controller performance analysis for 20 MW change in active power.

slightly higher value of 1st peak in dc power plots, following the change in power reference. For the same waveforms, the consecutive peaks (both upper and lower) attain relatively lower amplitude than the waveform from MMC without the tertiary controller. Following the change in active power, the d-axis component (I_{sd}) of the fundamental current also increases. The plots behave similarly to the active power plots. The q-axis component (I_{sq}) follows the reactive power, and the same is visible in the plots.

Change in an active power reference also causes the increase in circulating current, which is suppressed in a few cycles. The circulating current waveforms of all three MMCs are identical and do not show any observable impact of the tertiary controller. The i_z and P_z plots show the effectiveness of the controllers. For the MMC without tertiary controllers, the current and power oscillations are the same as in the previous case. They attain the amplitude up to 30 A and 15 MW, respectively. With PR and PI tertiary controllers, the oscillations are suppressed. On the instant of the change in reference power, the waveform of the MMC with PI controller shows a very small (almost negligible) increase. The transient condition causes no observable change in the waveform for the MMC with the PR controller.

5.7.3 Transient operation: 20 MVAr change in reactive power

Fig. 5.16, presents the plots for change in reactive power operation. Initially, the MMCs are operating at the rated voltage with 400 MW active power and 0 MVAr reactive power support. At 1.8 seconds, reactive power has been increased by 20 MVAr (Fig. 5.16). Following the change in reference reactive power, the reactive power support from all three MMCs increases instantly. Unlike the P_{ac} waveforms of the previous case (Fig. 5.15), all three waveforms attain the same peak value at the instant of the change. For the P_{ac} plot, the amplitude of the oscillation is minimum when the tertiary controller is not applied and it is maximum with the PI controller. It is in contrast to the previous case (Fig. 5.15), where oscillation in Q_{ac} is identical. Similar to the previous case, the current I_{sd} and I_{sq} follow the active power and reactive power plots, respectively. The I_{sq} plot witnesses the step-change following the change in reactive power. The I_{sd} plot observes fluctuation due to the coupling effect.

Due to the coupling with active power, dc power also witnesses the oscillations, which die out gradually. Initially, the amplitudes of oscillations for the MMCs without a controller and PI-based controller are the same, while slightly (negligible) lower in the case of the PR controller. In terms of circulating current, all three MMCs present identical plots. In an absence of a tertiary controller, the i_z plot

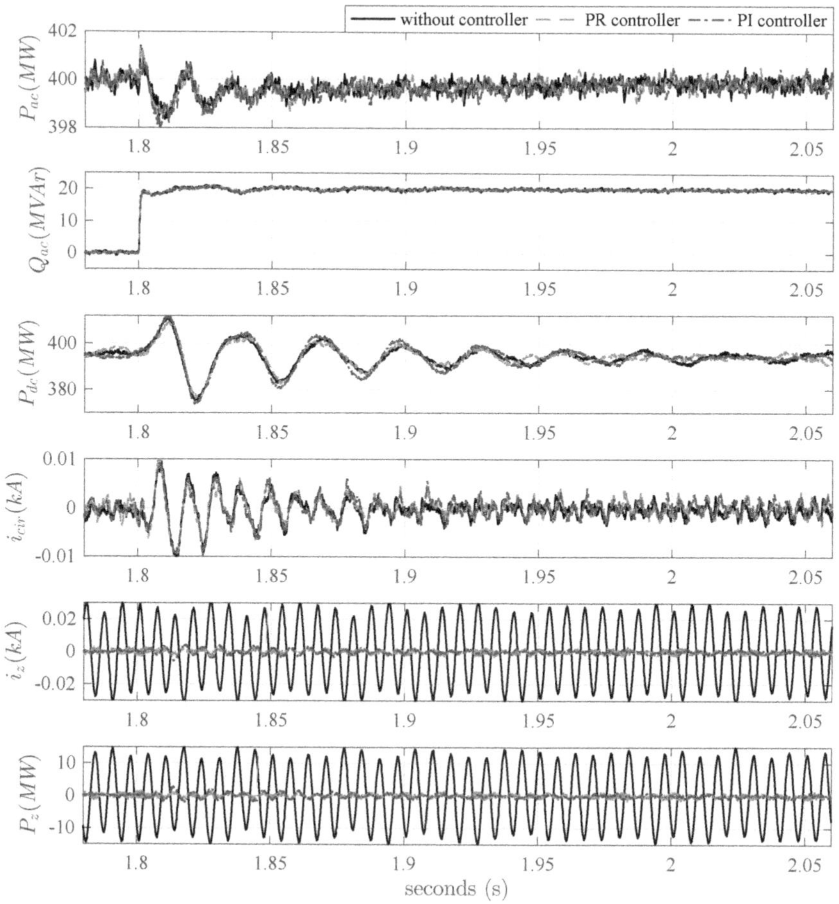

Fig. 5.16. Controller performance analysis for 20 MVAr change in reactive power.

reaches up to 30 A as in previous cases. With controllers, this current is suppressed successfully. The P_z plot follows the i_z plot with a constant dc voltage.

Based on the analysis of plots, it can be summarised that both controllers (PI and PR) are equally efficient in suppressing the 3rd order zero-sequence current flowing through transformerless MMC arms and source. It in turn suppresses the power oscillations on the dc lines. The analysis of active and reactive power plots under transient conditions shows no effect on the reactive power in the presence of the tertiary controller. However, the active power flow is slightly affected.

5.7.4 Impact on internal dynamics

Fig. 5.17 presents the IHD of the 3rd order harmonic current under steady-state operation. It is 1.4% of the fundamental current. With the controller, it is almost zero. Further, the performance of the PI controller is slightly better than the PR controller. Fig. 5.18 analyses the impact of tertiary controllers on the internal dynamics of the transformerless MMC. The arm current and SM capacitor voltage are considered for the study.

If the controller does not suppress the 3rd order zero-sequence current (i_z), it attains a value up to 30A. However, as shown in the I_s plot, the elimination of I_z

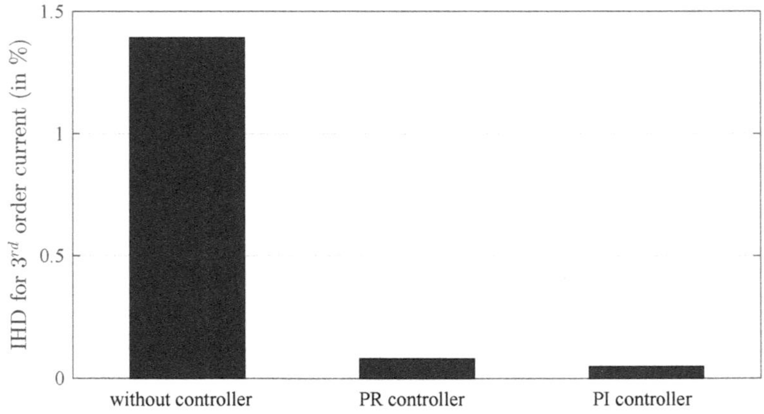

Fig. 5.17. IHD of 3rd order zero-sequence harmonic current in phase 'a'

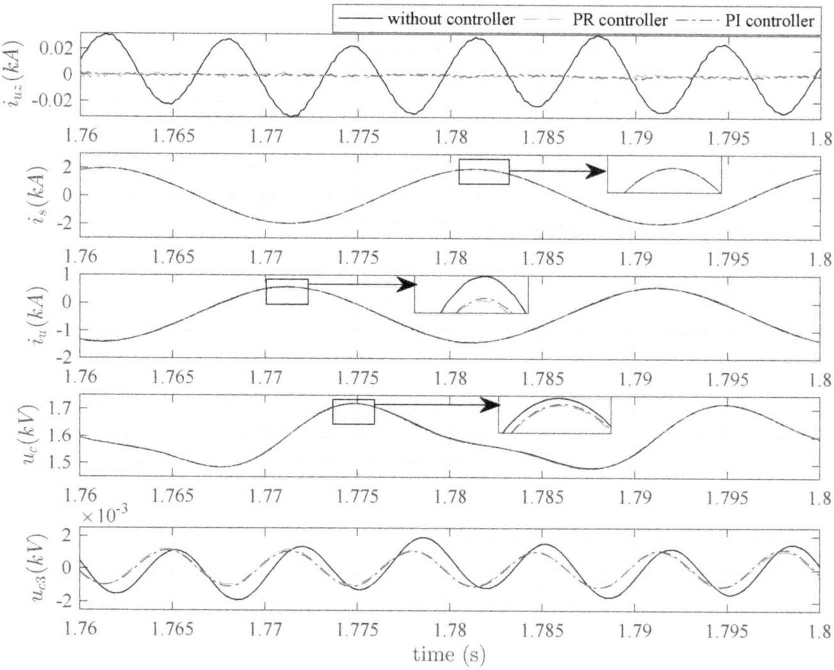

Fig. 5.18. Impact on internal dynamics

does not affect the fundamental current. This means the presence of the I_z does not cause any adverse effect on the fundamental current. As each 3rd peak of the zero-sequence harmonic current coincides with the arm current peaks, its presence increases the amplitude of the arm current. The elimination of the 3rd harmonic current reduces the peak value. Since the current i_z is inversely proportional to the arm inductance, its impact could be significantly large in MMCs with low arm inductance. Both controllers give comparatively the same degree of reduction in the peaks. In the case of SM capacitor voltage u_c, the elimination of zero-sequence current leads to a reduction in peaks, however, it is negligible. Additionally, the controllers cause a phase shift too. The interaction of the dc, fundamental, second, and higher-order currents and SM capacitor Voltage cause the ripples of harmonic voltage, including the third-order component. The tertiary controller eliminates the third-order current by generating a third-order voltage in the SM capacitors.

The cumulative effect of the third-order voltage generated due to tertiary controller action and the third-order voltage due to the dc, fundamental and other components causes the phase shift and peak reduction. Further, it can be summarised that the tertiary controllers do not cause any adverse impact on the internal dynamics of the system.

5.8 Conclusion

This chapter has successfully analysed the transformerless MMC using the real-time digital simulator (RTDS). In the transformerless MMC, with mid-point grounded dc terminals, the zero-sequence current flows through the MMC as well as the ac source and can cause serious damage. The dynamic equations have been updated to include the transformerless mode. PR and PI controllers have been designed to eliminate the unwanted third-order current. The bode plot shows that both PI and PR controllers are stable. A 22nd order state-space model has been developed for the transformerless MMC with the tertiary PI controller. The analysis of this model asserts the system's stability. The implementation of tertiary controllers enhances the stability of the system. The performance of the PI controller has been compared with the PR controller. Both steady state and dynamic analyses demonstrate the effectiveness of both controllers in suppressing the third-order current and the controller does not affect any internal or external variable adversely. With the proposed configuration, the utilities can significantly minimise the converter station cost.

Chapter 6

Design and Analysis of Symmetrical Multi-pole MMC HVDC

6.1 Introduction

For long-distance power transfer, high voltage dc (HVDC) is the best choice. With the significant progress in the converter technologies, specially voltage source converter (VSC), the HVDCs are being installed throughout the globe. The modular multi-level converter has proven to be the most advanced VSCs. The recent advancement in the HVDC breaker technology [144, 145] brings additional momentum to MMC HVDC implementation. The recently developed HVDC breakers are able to break up to 20 kA current in 0.003 seconds [10, 146, 147]. Many existing high voltage ac transmission lines can be upgraded to the HVDC network using the proposed multipolar configuration. It can be done by installing MMCs at sending and receiving ends. The existing healthy ac transmission conductors can be used for HVDC transmission, minimising the investment on the line. Since the modern ac transmission lines are mostly three-phase systems, and for upgrading them into an HVDC transmission line, a three-wire-based system would be more suitable.

The available literature presents the basic idea for MMC multi-pole configuration. However, a detailed analysis is to be carried out. In [84], authors have proposed a half-bridge submodule (HBSM) and full-bridge submodule (FBSM) based tri-pole HVDC. The study uses HBSM-based MMC (H-MMC) and FBSM-based MMC (F-MMC). However, the F-MMC is not utilised to full capacity despite being the most costlier than the H-MMC from the economic perspective. The use of the F-MMC pole is limited to the supporting role only. Further, the fault analysis is also not considered in the analysis. The concept of symmetrical and asymmetrical multi-pole MMC HVDC has been introduced in [85]. It also analyses the operation of tri-pole under faults. The proposed solution reduces the power transfer capacity of the network. Moreover, the effect of one MMC on other under a multi-pole configuration has not been considered by either one.

A fault-type discrimination methodology using the rate of change of current is proposed in [148]. In [149], authors have presented a support vector machine regression mechanism for determining the fault location of the single pole-to-ground. However, the method to isolate the fault and recovery after the fault is cleared is missing. A protection scheme has been developed in [150], but is limited only to isolating the faulty section. In [151], authors propose the active current-limiting control for both ac and dc faults. The dc pole to ground fault ride capability is proposed in [152]. However, for the same, the authors have added an additional switch in each leg of the MMC. Moreover, the dc line to line (LL) fault is not considered in the study. In [153], authors have analysed the system behaviour under pole-to-pole fault. Since, the HBSM lacks the dc fault handling capability, they have used a hybrid structure for the same.

In the present chapter, first, the analysis of H-MMC-based symmetrical and asymmetrical multi-pole has been carried out. One terminal of the HVDC is realised using H-MMC, while the other one is a constant dc voltage source. Next, the symmetrical multi-pole MMC-HVDC system has been investigated in detail. Both the terminals have been realised using H-MMC only. It has been further divided into even and odd symmetrical multi-pole. Mathematical relations between pole variables are included. A controller with fault-ride-through capability under dc side faults has been proposed for the tri-pole version of the symmetrical multi-pole. The analysis of the control structure of MMC HVDC reveals that the controller output is dependent on dc line current and voltages measurements. During fault conditions, it affects the system adversely. The proposed controller has digital switches to temporally bypass the measurement of faulty values by measurements taken under healthy conditions. The performance of the proposed controller has been carried out on a real-time digital simulator (RTDS).

The main contributions of this chapter are as follows.

1 Design and analysis of three-wire symmetrical and asymmetrical multi-pole configuration.

2 Design of a symmetrical multi-pole MMC HVDC configuration and derivation of the corresponding mathematical relationships between the variables.

3 Impact analysis on the entire system following the change in the operating condition of one MMC.

4 The controller and its control algorithm for dc fault ride-through are developed, and its effectiveness is demonstrated through the simulations on RTDS.

6.2 Multipolar MMC HVDC Configurations

The absence of a solid ground connection of dc side ensures that MMC can operate in unipolar and bipolar modes. Under bipolar mode, the voltage levels are $+V_{dc}/2$ and $-V_{dc}/2$ for upper and lower poles, respectively. However, with unipolar mode, one pole is at 0 kV while the other one is either at $+V_{dc}$ or $-V_{dc}$, ensuring V_{dc} voltage between two dc poles.

The modular structure of MMC infuses it with the advantage of scalability. This, however, is not a sufficient requirement for upgrading an ac transmission network

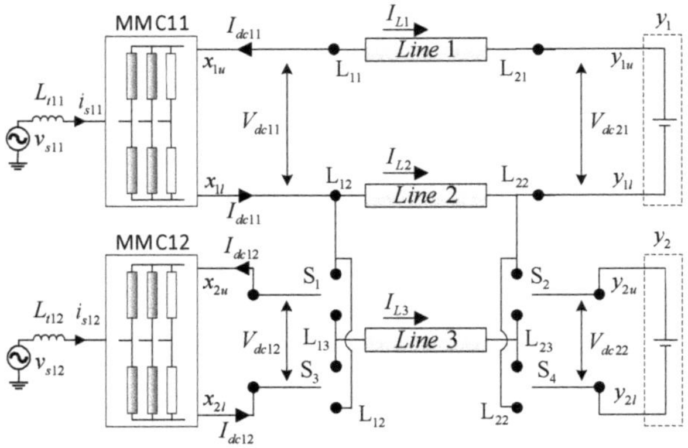

Fig. 6.1. Multipolar configuration

into a dc one. A typical MMC HVDC configuration utilises only two transmission lines, while the ac networks consist of a minimum of three lines. By combining two MMCs, a multipolar configuration can be achieved, which uses the three transmission lines and can be used for upgrading existing ac transmission lines to dc transmission lines. Two such multipolar configurations for MMC HVDC have been developed and presented in Fig. 6.1. One terminal of HVDC is made of H-MMC, while the other is represented by dc source. The two MMCs are connected to two dc sources via three transmission lines. With the proper connection of switches, symmetrical and asymmetrical multi-pole configurations can be derived. The switch positions are explained in Table 6.1. The two ends of ith transmission line at the terminal T are represented by L_{Ti}.

Table 6.1: Switch positions for SYMC and ASYMC configurations

Switch	S1	S2	S3	S4
SYMC	L_{12}	L_{22}	L_{13}	L_{23}
ASYMC	L_{13}	L_{23}	L_{12}	L_{22}

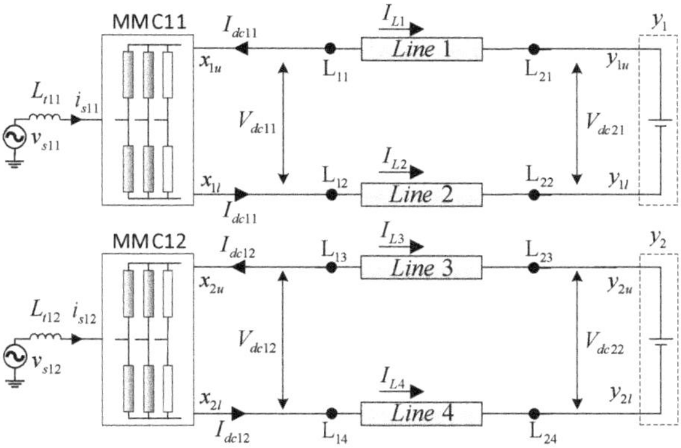

Fig. 6.2. Base configuration: Parallel MMC HVDC

6.2.1 Base configuration

The base configuration is the primary of two MMC HVDC systems operating independently, as depicted in Fig. 6.2. MMC11 is connected to dc source y_1 between lines 1 and 2, and similarly, MMC12 is connected to y_2 between lines 3 and 4. The MMCs are operating in bipolar mode with the terminal dc voltage at V_{dc11} and V_{dc12}. The upper and lower dc poles of MMC11, i.e. x_{1u} and x_{1l}, are at $+V_{dc11}/2$ and $-V_{dc11}/2$, respectively. The currents through lines 1 and 2 are the same as the current from upper and lower dc poles, i.e. $+I_{dc11}$ and $-I_{dc11}$. Similarly, the for MMC12, dc poles x_{2u} and x_{2l}, are at $+V_{dc12}/2$ and $-V_{dc12}/2$. Line 3 is connected to x_{2u} and y_{2u}, and line 4 is connected to x_{2l} and y_{2l}, carrying current $+I_{dc12}$ and $-I_{dc12}$, respectively.

6.2.2 Symmetrical multipolar configuration (SYMC)

Table 6.1 presents the switch status for the two configurations. When switches are operated in such a way that the lower dc pole of MMC11 (x_{1l}) is connected to the upper dc pole of MMC12 (x_{2u}), we get a symmetrical configuration (SYMC) as depicted in Fig. 6.3. Pole x_{2u} is connected to pole x_{1l} and line 2 through switch S_1.

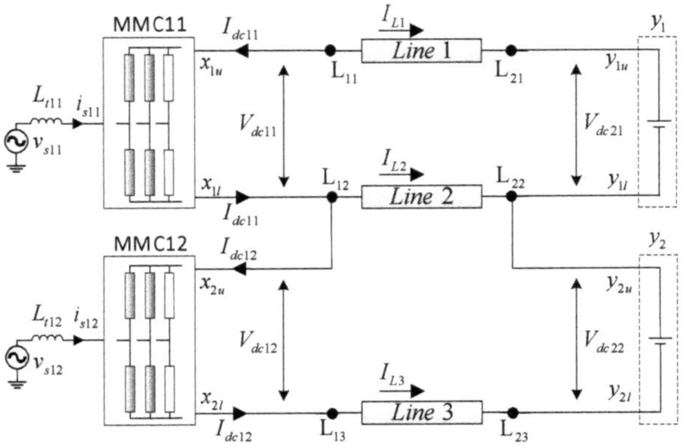

Fig. 6.3. Symmetrical multipolar configuration (SYMC)

At the other end of line 2, switch S_2 connects pole y_{1l} to pole y_{2u}. Line 3 is connected between poles x_{2l} and y_{2l} through switches S_3 and S_4. Under this configuration, line 2 carries the current for both MMCs, which can be defined as in (6.1). Line 1 and line 3 carry the currents of the respective poles, i.e. $-I_{dc11}$ and I_{dc12}, respectively. If both MMCs are identical and supply the same amount of power, then the current in I_{L2} becomes zero as $I_{dc11} = I_{dc112}$.

$$I_{L1} = -I_{dc11}; \quad I_{L2} = I_{dc11} - I_{dc12}; \quad I_{L3} = I_{dc12} \tag{6.1}$$

$$V_{x_{1l}} = V_{x_{2u}} = V_{L_{12}} \tag{6.2}$$

$$V_{x_{1u}} = V_{L_{12}} + V_{dc11}; \qquad V_{x_{l2}} = V_{L_{21}} - V_{dc12} \tag{6.3}$$

$V_{L_{12}}$ represents the voltage at the junction of both MMCs with line 2 as in (6.2). The MMC HVDC maintains the reference voltages between dc poles, and thus, the voltage at the upper dc pole MMC11 and lower dc pole of MMC12 (i.e. $V_{x_{1u}}$ and $V_{x_{2l}}$, respectively) are defined by (6.3). With identical MMCs and equal dc terminal voltages (i.e. $V_{dc11} = V_{dc12}$), junction voltage $V_{L_{12}}$ attains a zero value, forcing MMCs to operate in unipolar mode. However, when $V_{dc11} \neq V_{dc12}$, $V_{L_{12}}$ attains a non-zero value.

6.2.3 Asymmetrical multipolar configuration (ASYMC)

In an asymmetrical configuration (ASYMC), the lower dc pole of MMC11 is connected to the lower dc pole of MMC12 as shown in Fig. 6.4. Both poles (i.e. x_{1l} and x_{2l}), and line 2 are connected by switch S_3. At the other end, line 2 is connected to lower pole of dc sources (i.e. y_{1l} and y_{2l}) by switch S_4. Switches S_1 and S_2 connect line 3 to the upper pole of MMC12 and dc sources y_2, i.e. x_{2u} and y_{2u}, respectively. In this configuration, line 2 carries the current of both MMCs and is defined in (6.4). Currents in line 1 and line 3 are $-I_{dc11}$ and $-I_{dc12}$, respectively. The voltage for the lower poles can be defined as in (6.5), which is an average of negative pole voltages under base configuration. The MMC maintains the dc voltage at the given pole-to-pole reference. This may lead to a shift in upper pole voltages as in (6.6), in case of $V_{dc11} \neq I_{dc12}$.

$$I_{L1} = -I_{dc11}; \qquad I_{L2} = I_{dc1} + I_{dc2} \qquad I_{L3} = -I_{dc12} \tag{6.4}$$

$$V_{x_{1l}} = V_{x_{2l}} = V_{L_{12}} = -(V_{dc11} + V_{dc12})/4 \tag{6.5}$$

$$V_{x_{1u}} = V_{L_{12}} + V_{dc11}; \qquad V_{x_{2u}} = V_{L_{12}} + V_{dc12} \tag{6.6}$$

Thus, the two multipolar configurations using two MMCs can be obtained. With identical MMCs and operating conditions, in SYMC, the upper dc pole of MMC11 and lower dc pole of MMC12 are under twice voltage stress than in the base case, while the other pole is at zero voltage. Current in line 2 is zero and in lines 1 and 3, it attains the nominal value. In ASYMC, the dc poles are under nominal voltage stress, i.e. either at $+V_{dc}/2$ or $-V_{dc}/2$, but line 2 has to carry a current twice that of line 1 or 3. Table 6.2 shows voltage and current values for different configurations. When the MMC are operating under unequal dc voltages, the junction point of two MMCs in ASYMC attains a voltage level which is the average of negative pole voltages of both MMCs, if they are operating independently. However, in SYMC,

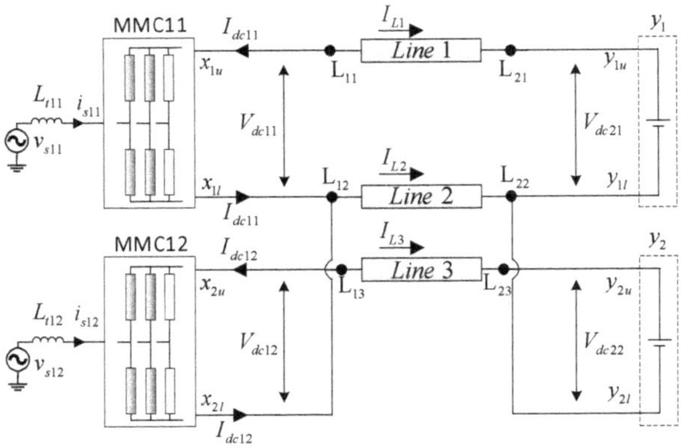

Fig. 6.4. Asymmetrical multipolar configuration (ASYMC)

the junction point, which is at zero voltage for an identical case, attains a non-zero value under an unequal voltage scenario.

Table 6.2: Pole voltages and line currents for multipolar MMC HVDC

Parameters	Base Configuration	SYMC	ASYMC
$V_{x_{1u}}$	$V_{dc11}/2$	$V_{L_{12}} + V_{dc11}$	$V_{L_{12}} + V_{dc11}$
$V_{x_{1l}}$	$-V_{dc11}/2$	$V_{L_{12}}$	$V_{L_{12}}$
$V_{x_{2u}}$	$V_{dc12}/2$	$V_{L_{12}}$	$V_{L_{12}} + V_{dc12}$
$V_{x_{2l}}$	$-V_{dc12}/2$	$V_{L_{12}} - V_{dc12}$	$V_{L_{12}}$
I_{L1}	$-I_{dc11}$	$-I_{dc11}$	$-I_{dc11}$
I_{L2}	I_{dc11}	$I_{dc11} - I_{dc12}$	$I_{dc11} + I_{dc12}$
I_{L3}	$-I_{dc12}$	I_{dc12}	$-I_{dc12}$

The main advantage of the SYMC is the zero current in line 2, which also means zero losses. However, the twice voltage stress at the outer poles can be considered as its limitation as it increases the insulation requirements. In ASYMC, the voltage stress is the same at each pole of each MMC. However, the current in line 2 is twice that of other lines, leading to excessive power loss. In terms of installation cost, both configurations stand equal, as they use identical MMC units. However, the power loss in line 2 of ASYMC makes it less cost-effective than SYMC during operation. Though, either of the configurations can be installed independently, the switches' presence gives an additional dimension to operation. Now the system can be switched between the two modes if required. The addition of the switches would increase the installation cost. But it would be negligible in comparison to the cost of the converter station.

6.3 Performance Analysis of SYMC and ASYMC

For the performance analysis of the proposed configurations, the simulations have been performed on the RTDS platform. MMC11 operating conditions are kept constant during the analysis. For MMC12, operating parameters have been varied in accordance with the test case. The three test case scenarios considered are 1) change in active power (P_{ac}), 2) change in reactive power (Q_{ac}), and 3) change in dc voltage (V_{dc}). Both MMCs are designed for 500 MVA, 166 kV ac voltage (line-to-line) and 320 kV dc side voltage (pole-to-pole). An arm of MMC consists of 200 SMs, each rated at 1.6 kV. Initially, at $t = 0$, both MMCs are operating with rated voltages, 450 MW active power and 0 MVAr reactive power. At the time $t = 0.2$ second, the operating parameters of MMC12 are set to new values, and the response of both MMCs, as well as that of dc lines, are recorded. The results are presented in Figs. 6.5-6.10. The solid line represents the MMC12 parameters, while the dashed line represents the MMC11 parameters. A dash-dot line depicts the current in dc lines.

6.3.1 With identical operating conditions

As mentioned above and shown in Figs. 6.5 and 6.6, initially both MMCs are operating at 450 MW, 0 MVAr and rated voltages. Since both are handling equal power flow, the current I_{dc} is equal for both, i.e. -1.4 kA. Further, for SYMC, I_{dc11} represents $-I_{L1}$ and I_{dc12} represents I_{L3}. Thus currents in lines 1 and 3 are of equal magnitude (i.e. 1.4 kA) but in opposite directions. Current in line 2 (i.e. I_{L2}), which is equal to the difference of I_{dc11} and I_{dc12}, is at zero, as shown in the plot.

The V_{x1l} (and V_{x2u}) represents the voltage to the junction point of both MMCs, and it is at 0 V. The other pole voltages, i.e., V_{x1u} and V_{x2l} are at 320 kV and -320 kV, respectively. It means that the MMCs are operating under unipolar conditions. Similarly, in ASYMC, I_{dc11} and I_{dc12} represent $-I_{L1}$ and $-I_{L3}$, respectively. Thus, the currents in lines 1 and 3 are of the same magnitude (i.e. 1.4 kA) and direction. Here, current I_{L2} is the sum of currents in the other two lines and attains a value of -2.8 kA. The upper and the lower pole voltages are at 160 kV and -160 kV, respectively, leading to the bipolar mode of operation with dc terminals virtually grounded at the midpoint. For both configurations, SM capacitor (SM capacitor) voltages of both MMCs are overlapping. SM capacitor voltages are oscillating around its dc component, i.e. 1.6 kV. Further, The circulating currents, i_{cir}, are zero.

6.3.2 Change in active power from 450 MW to 500 MW

Figs. 6.5 and 6.6, respectively, present the plots for SYMC and ASYMC, under the active power change. For both configurations, at the instant t = 0.2 second, the active power flow of MMC12 is increased by 50 MW. The P_{ac} plots show that, the MMC12 attains the reference value of 500 MW within two cycles. However, the coupling effect causes the fluctuations in Q_{ac} plots of MMC12. For MMC11, the active and reactive plots remain unaffected. The MMC12 also witnesses a sudden increase and elimination of circulating currents, associated with fluctuations in Q_{ac}. An increase in circulating current leads to an increase in the arm current and associated losses. For both SYMC and ASYMC, the SM capacitor voltages u_c show no variations at all. The dc pole voltages V_{xu} and V_{xl} witness a swing of negligible magnitude. For SYMC, as depicted in the zoomed box, it is more observable in the poles connected to line 2, i.e. x_{1l} and x_{2u}. In the case of SYMC, only the pole voltage of MMC12 is affected. The increase in power flow increases the dc current of MMC12. Thus, I_{dc12} is no longer equal to I_{dc11}. This means, for SYMC, $I_{L1} \neq I_{L3}$ and also I_{L2} is no longer at zero and attains a value of $\approx 1.56kA$. Similarly, for ASYMC $I_{L1} \neq I_{L3}$ as I_{dc12} has attained a new value. The current in line 2 (I_{L2})

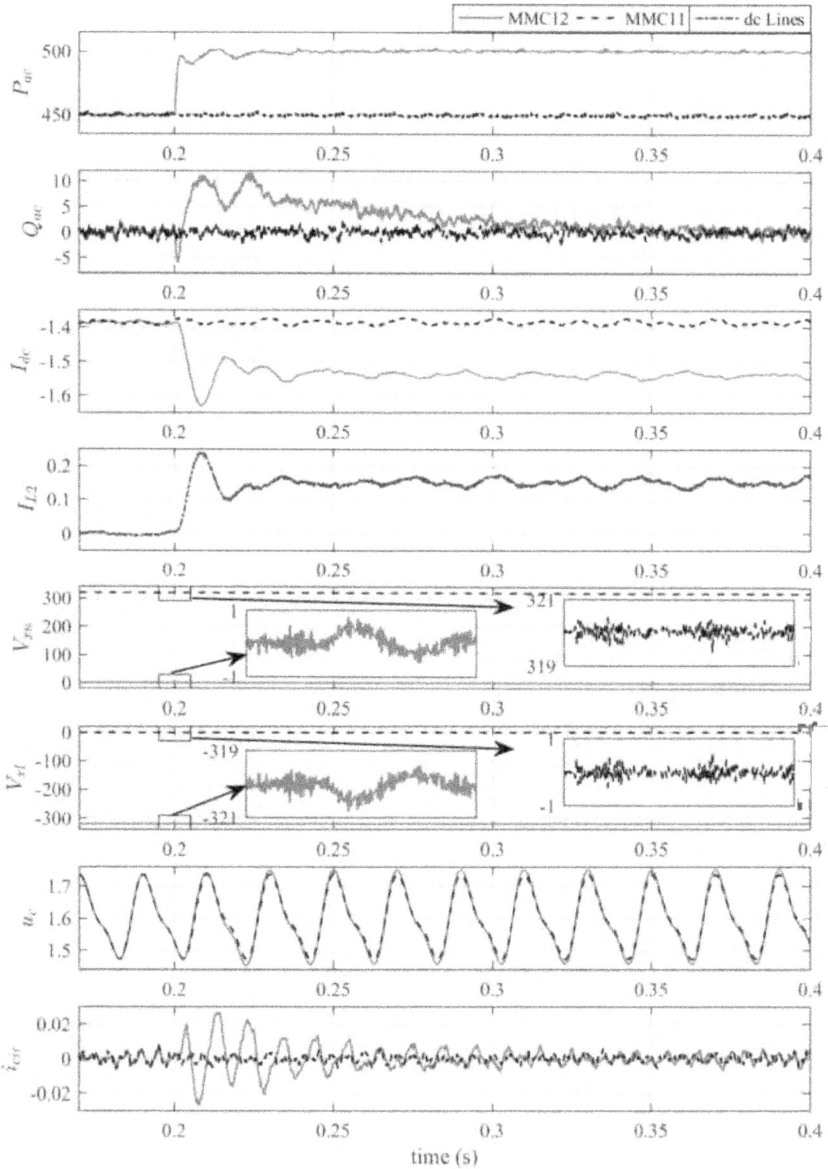

Fig. 6.5. Performance analysis of SYMC under the change in the active power of MMC12

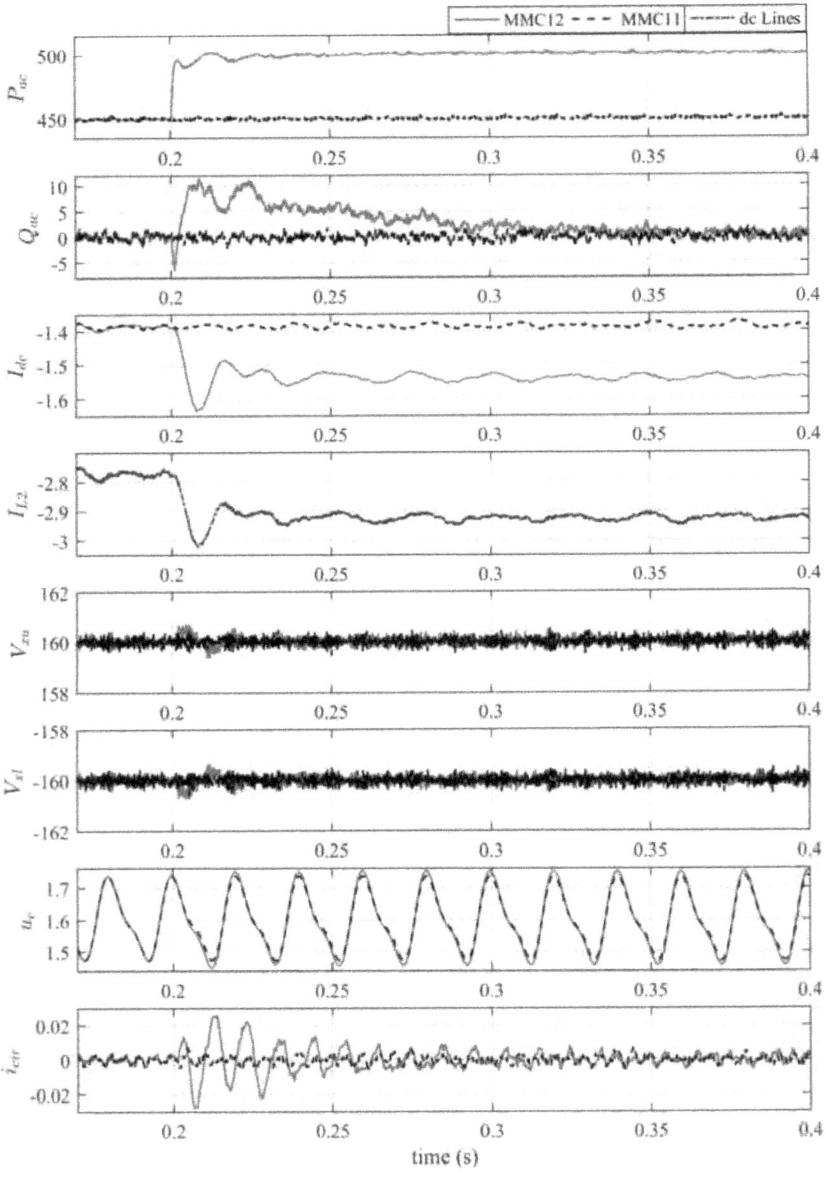

Fig. 6.6. Performance analysis of ASYMC under the change in the active power of MMC12

attains a value of ≈ -2.96 kA.

It can be summarised that all these variations are entirely related to MMC12 only. For both the configurations, MMC11 variables are entirely insensitive to active power change in MMC12.

6.3.3 Change in reactive power from 0 MVAr to 50 MVAr

For analysis of change in reactive power, the operating conditions are reset to 400 MW, 0 MVAr and rated voltages. At $t = 0.2$ second, the reactive power of MMC12 is increased by 50 MVAr. The plots are presented in Figs. 6.7 and 6.8 for SYMC and ASYMC, respectively. For both configurations, the change in the reactive power also causes fluctuation in the active power. Similar to the previous case, the impact on pole voltages is negligible. The SM capacitor voltages of both MMCs overlap and are not affected by the changes. The plots of dc current differ from that of the previous case. The dc current I_{dc12} shows the ripples under transient conditions and re-attains its previous values. It is linked to the coupling between reactive and active power. The effect is propagated to line 2 as well, as it carries the current for both the MMCs. Here also, MMC11 variables are insensitive to reactive power change in MMC12.

6.3.4 Change in dc voltage from 320 kV to 300kV

Similar to previous cases, for a change in dc voltage analysis, the MMCs are reset to 400 MW, 0 MVAr, and rated voltages, i.e. 166 kV line-to-line ac and 320 kV pole-to-pole dc voltage. At time instant $t = 0.2$ second, the dc voltage of MMC12 is reduced from rated 320 kV to 300 kV. The plots are presented in Fig. 6.9 and Fig. 6.10 for SYMC and ASYMC, respectively.

The most significant impact of voltage change can be seen at dc poles. Following the change in voltage reference of MMC12, while the pole-to-pole voltage attains a new reference in less than two milliseconds, the dc poles of both MMCs observe a drag for almost seven milliseconds before attaining the new equilibrium point.

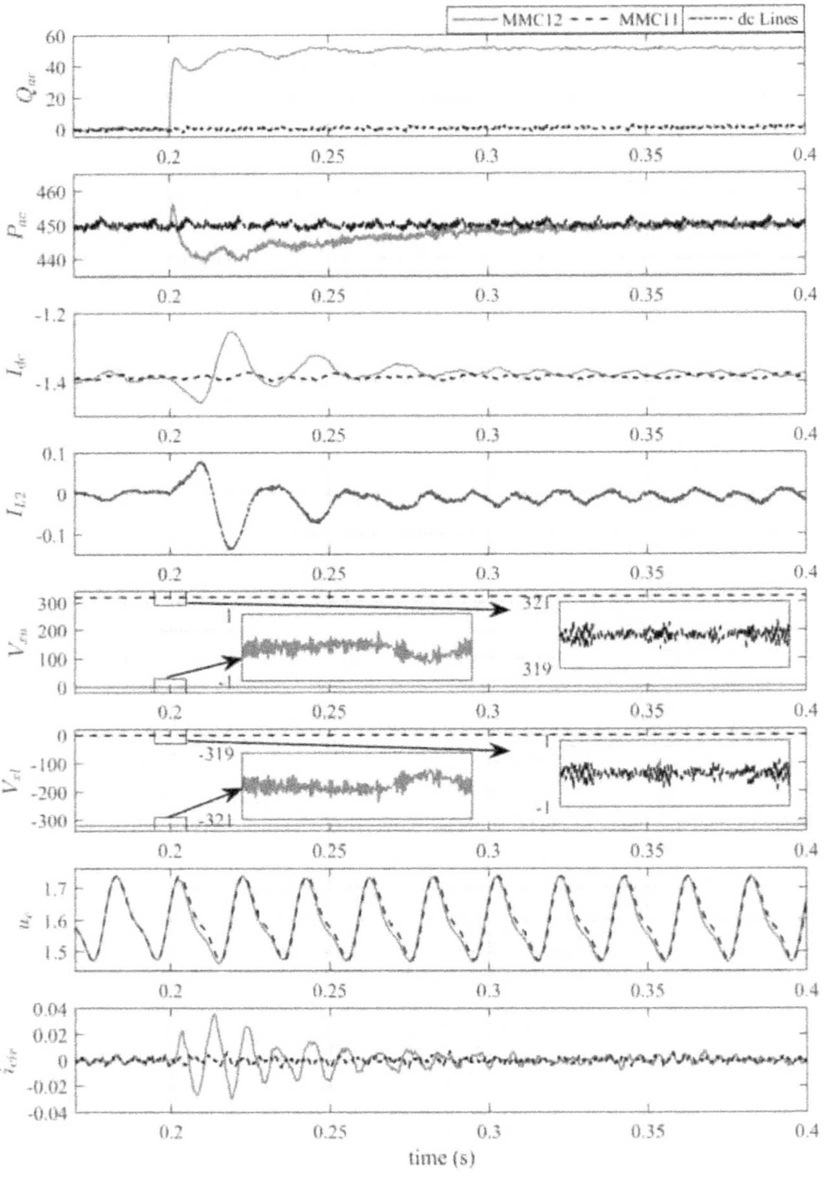

Fig. 6.7. Performance analysis of SYMC under the change in reactive power of MMC12

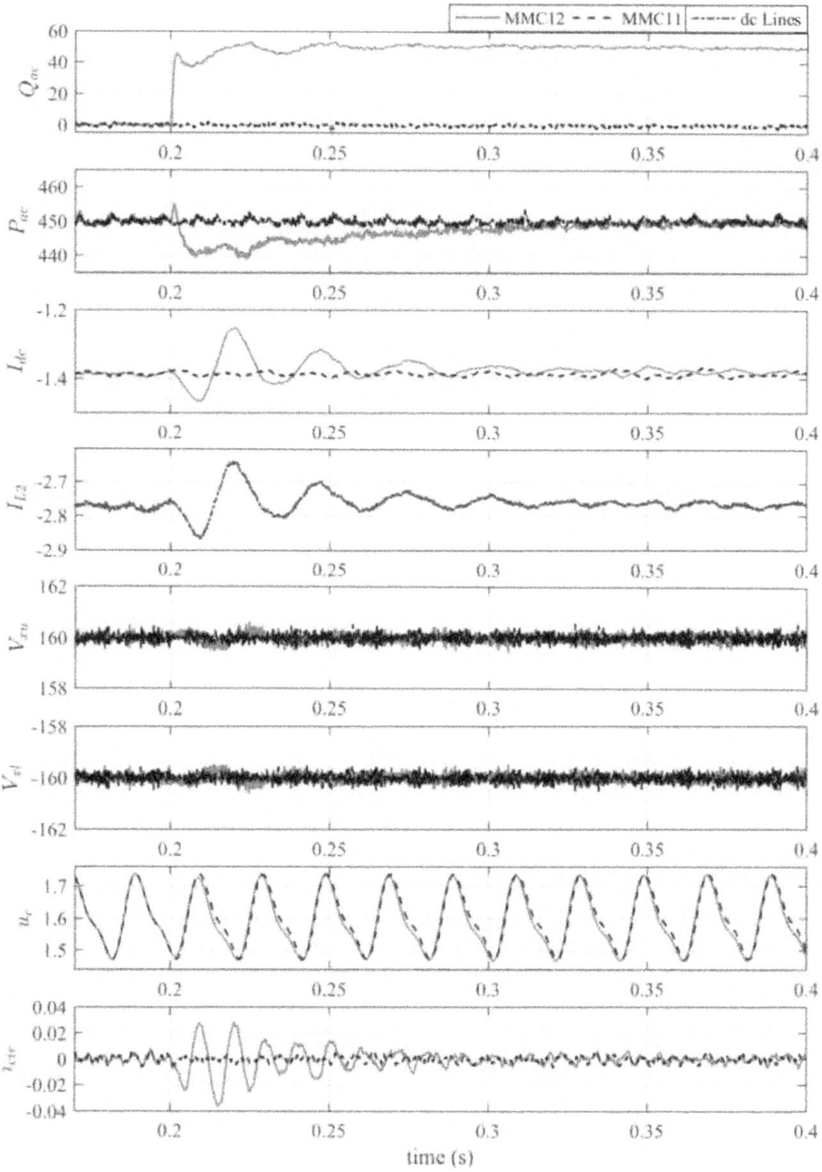

Fig. 6.8. Performance analysis of ASYMC under change in reactive power of MMC12

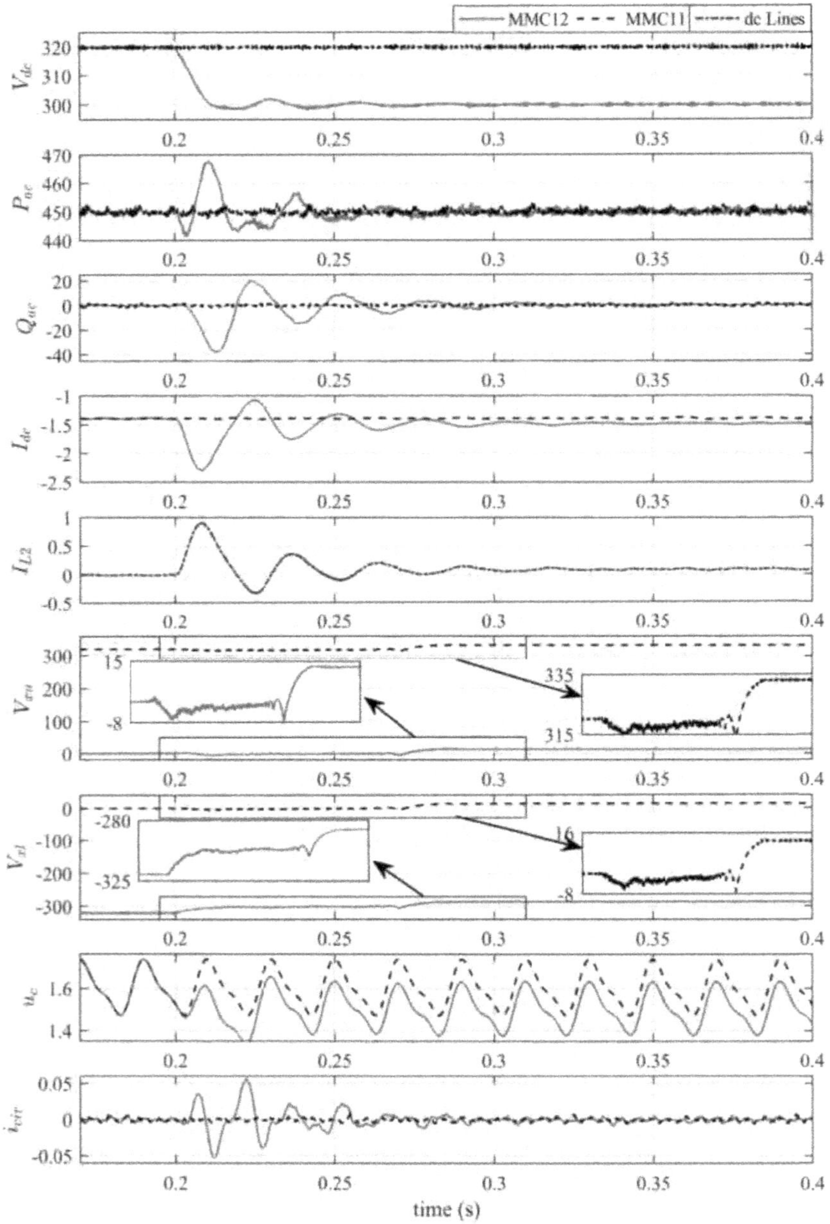

Fig. 6.9. Performance analysis of SYMC under the change in voltage

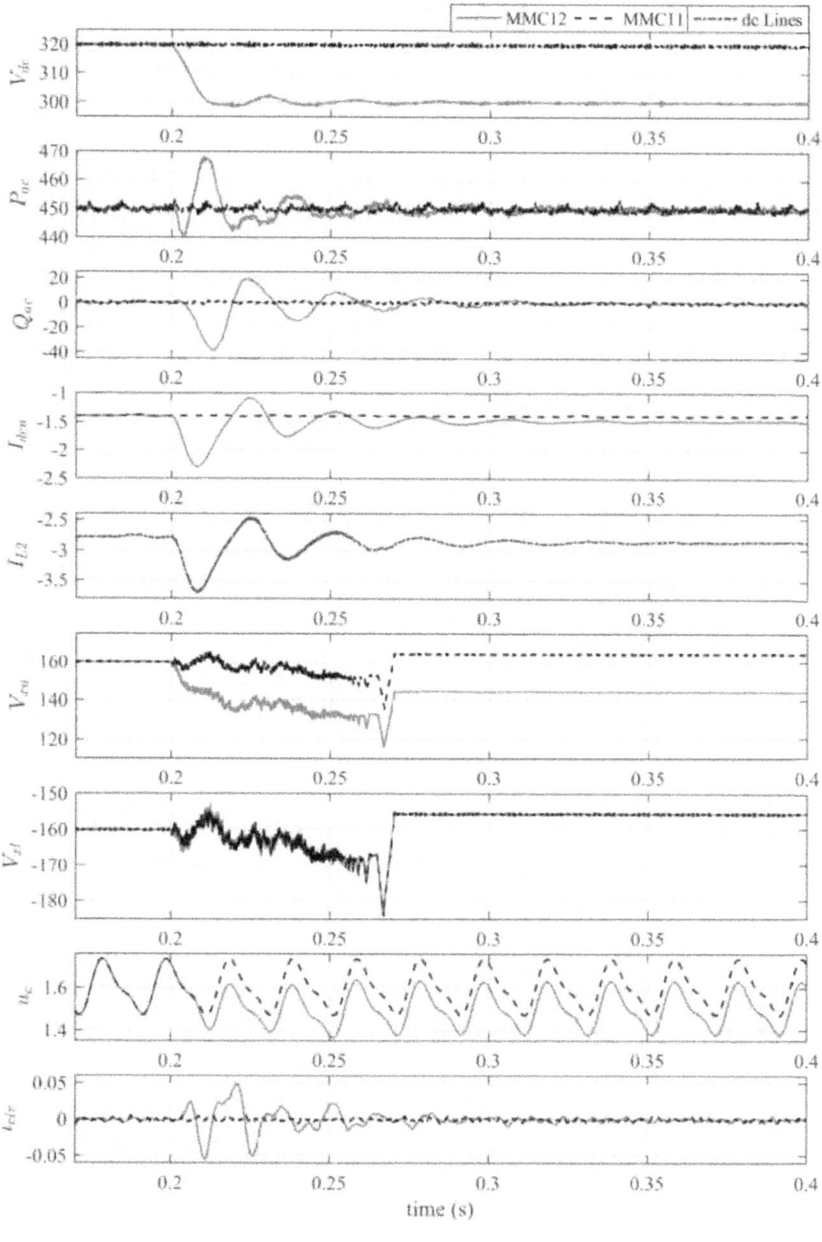

Fig. 6.10. Performance analysis of ASYMC under the change in voltage

For ASYMC, initially, the negative poles of both MMCs are at -160 kV, and the upper poles are at +160 kV. In accordance with (6.5) and (6.6), they settle to new values, which is -155 kV, for the lower pole of both MMCs, and 165 kV and 145 kV, respectively, for the upper pole of MMC11 and MMC12. Similarly, in SYMC, the junction point of both MMCs, which is initially at zero, post-drag settles to 13 kV. The new voltage level of the MMC11 upper pole is 333 kV, while the same for the MMC12 lower pole is -287 kV.

Further, in both configurations, the change in dc voltage results in ripples in active as well as reactive powers. For the change of 20 kV, the active power observes fluctuation, which reaches the peak value of 465 MW, and similarly, reactive power fluctuation reaches as high as 40 MVAr. SM capacitor voltage of MMC12, which is initially overlapping with the same for MMC11, reduces to a new average value of 1.5 kV. Following the voltage change, the circulating current of MMC12 increases and then attenuates as soon as active and reactive power fluctuations die out. The dc current I_{dc12} also witnesses the oscillation and settles to the higher value i.e. -1.5 kA, as the voltage reduces, but the power reference remains constant. Hence, I_{L3} is now -1.5 kA and 1.5 kA for SYMC and ASYMC, respectively, while I_{L1}, remains at +1.4 kA. The I_{L2} settles to 0.1 kA and -2.9 kA for SYMC and ASYMC, respectively.

Similar to previous cases, most of the MMC11 variables remain insensitive for change in dc voltage of MMC12. The same is not true for dc pole voltages. Instead, they are severely affected by the change. The most significant issue is that the reduction in dc pole-to-pole voltages of one MMC leads to an increase in the upper pole voltage of other MMC, as it shifts the reference point.

6.4 Symmetrical Multi-pole MMC HVDC

As explained in the previous section, the MMC can be arranged in symmetrical, asymmetrical or parallel configurations. This section explores the symmetrical multi-pole MMC HVDC configuration. Further, unlike in the previous section, here, both terminals of the HVDC network are formed by identical MMCs. Fig. 6.11

presents an architecture of symmetrical multi-pole. The MMCs are named "MMC Ti", where T is '1' or '2', respectively, for terminal 1 and terminal 2, whereas 'i' stands for its sequence and can reach up to the maximum number of MMC (X) at a particular terminal. The dc voltage of the particular MMC is V_{dcTi}. Similarly, all other variables are also defined in accordance with the MMC. On the ac side, the MMCs can be connected to a dedicated (or common) ac source.

On dc lines, the lower pole of ith MMC is connected to upper pole of $(i+1)^{th}$ MMC, and they share the $(i+1)^{th}$ dc line. The junction voltage of both poles is V_{LTi+1}. Except for the two outer lines i.e. 1_{st} and $X+1^{th}$, the lines carry the dc current for two adjacent MMCs, which is primarily the difference current and can be defined as:

$$I_{Li(i\neq 1,X+1)} = -I_{dc1i} + I_{dc1i-1} = I_{dc2i} - I_{dc2i-1} \tag{6.7}$$

The total power carrying capacity of the symmetrical configuration can be de-

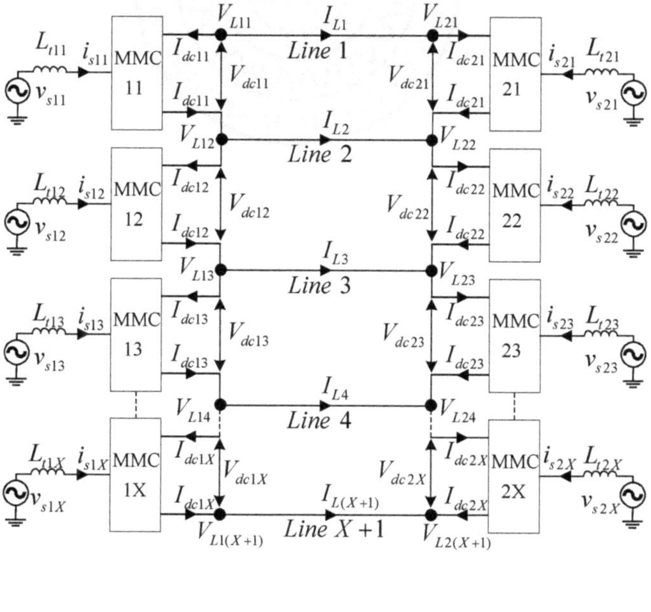

Fig. 6.11. Symmetrical multi-pole MMC HVDC configuration: basic architecture

fined as:

$$P_{Symmetrical} = \sum_{i=1}^{X} [V_{dcTi} * I_{dcTi}] \tag{6.8}$$

If the pole-to-pole voltages of all the MMCs at the terminal are equal (i.e. V_{dc}) and similarly current in all the lines are equal (i.e. I_{dc}), the total power of the symmetrical configuration can be defined as:

$$P_{Symmetrical} = XV_{dc}I_{dc} \tag{6.9}$$

With even and odd values of 'X', the configuration too can be termed as even or odd. Figs. 6.12 and 6.13 present the special cases of even and odd symmetrical multi-pole MMC.

6.4.1 Even symmetrical multi-pole MMC HVDC

Fig. 6.12 presents the even symmetrical multi-pole with X equal to two. It gives a tri-pole configuration. Line 1 and line 3 carry the load currents, while line 2 carries the difference current. Using (6.7), it can be derived as:

$$I_{L2} = -I_{dc12} + I_{dc11} = I_{dc22} - I_{dc21} \tag{6.10}$$

In the even symmetrical multi-pole, the physical middle pole acts as a mid-point. Its voltage V_{odcT} $(= V_{LTi(=\frac{X}{2}+1)})$ can be defined as the difference between the sum of the dc voltage of all MMCs above it, and the sum of dc voltage of all MMCs below it, as in (6.11).

$$V_{odcT} \approx \frac{1}{4}\left[\sum_{i=1}^{\frac{X}{2}} V_{dcTi} - \sum_{i=\frac{X}{2}+1}^{X} V_{dcTi}\right] \tag{6.11}$$

For the special case presented in Fig. 6.12, the mid-point voltage ($V_{odcT} = V_{LT2}$)

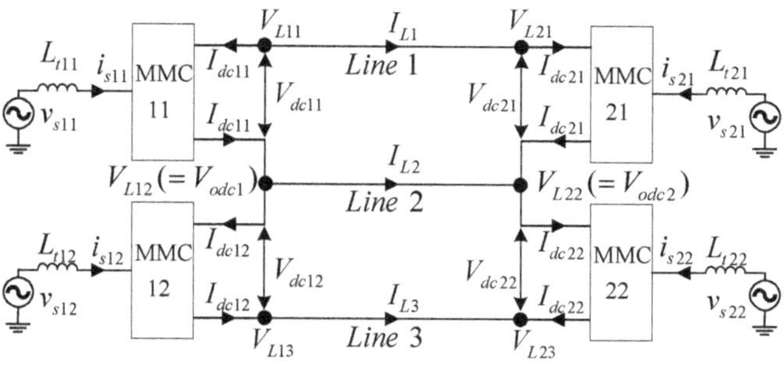

Fig. 6.12. Even symmetrical multi-pole MMC HVDC configurations: three pole architecture

can be derived as (6.12).

$$V_{odcT} \approx \frac{1}{4}\left[V_{dcT1} - V_{dcT2}\right] \tag{6.12}$$

With all the MMCs operating under identical dc voltage conditions, the V_{odcT} attains a zero value. However, it settles to a non-zero value if the voltages are nonidentical. All the other pole voltages can be deduced in reference to it.

6.4.2 Odd symmetrical multi-pole MMC HVDC

If the number of MMCs at each terminal is odd and more than one (i.e $X = odd(>1)$), it yields the odd configuration of multi-pole. A special case of an odd symmetrical multi-pole with three MMCs at each terminal ($X = 3$) is shown in Fig. 6.13. In this quad-pole configuration, only the outer lines (i.e. line 1 and line 4), carry the load current. While the inner lines (i.e. line 2 and line 3) carry the difference current only. Using (6.7), the current in these lines can be derived as:

$$\begin{aligned} I_{L2} &= -I_{dc12} + I_{dc11} = I_{dc22} - I_{dc21} \\ I_{L3} &= -I_{dc13} + I_{dc12} = I_{dc23} - I_{dc22} \end{aligned} \tag{6.13}$$

Unlike the even configuration, in the odd symmetrical multi-pole, the physical middle pole does not exist, instead, it has a middle MMC (at $i = (X+1)/2$). The dc side virtual midpoint voltage of this middle MMC can be defined as the mid-point voltage of the terminal T, V_{odcT} as in (6.14). Its upper and lower pole voltages (i.e. V_{LTi} and $V_{LT(i+1)}$), for $i = (X+1)/2$ are defined in (6.15). Using the same as reference, other pole voltages can be deduced.

$$V_{odcT} = \frac{1}{4}\left[\sum_{i=1}^{\frac{X-1}{2}} V_{dcTi} - \sum_{i=\frac{X+1}{2}+1}^{X} V_{dcTi}\right] \tag{6.14}$$

$$\begin{aligned} V_{LT\frac{X+1}{2}} &= V_{odcT} + \frac{1}{2}V_{dcT\frac{X+1}{2}} \\ V_{LT\frac{X+1}{2}+1} &= V_{odcT} - \frac{1}{2}V_{dcT\frac{X+1}{2}} \end{aligned} \tag{6.15}$$

For the quad-pole configuration of Fig. 6.13, the MMC12 and MMC22 are the middle MMCs of their respective terminals. From (6.14), the mid-point voltage can be derived as (6.16). With the outer MMCs operating under identical or nonidentical dc voltage conditions, the mid-point voltage attains a zero or non-zero value,

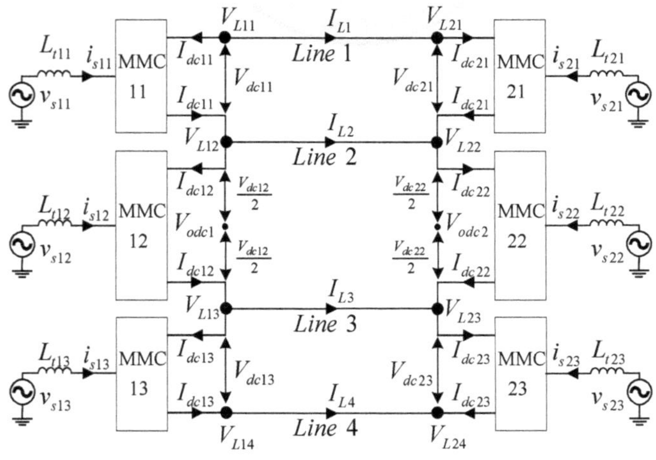

Fig. 6.13. Odd symmetrical multi-pole MMC HVDC configurations: four pole architecture

respectively.

$$V_{odcT} = \frac{1}{4}\left[V_{dcT1} - V_{dcT3}\right] \tag{6.16}$$

The pole voltages of the middle MMCs are deduced from (6.15) as given in (6.17).

$$\begin{aligned} V_{LT2} &= V_{odcT} + \frac{1}{2}V_{dcT2} \\ V_{LT3} &= V_{odcT} - \frac{1}{2}V_{dcT2} \end{aligned} \tag{6.17}$$

6.4.3 Retrofitting high voltage ac to high voltage dc

The symmetrical configuration can be employed to retrofit high voltage ac (HVAC) to HVDC. The usual HVAC transmission network is either a three-phase three-wire or a three-phase four-wire system. A tri-pole configuration can be used for retrofitting a three-phase three-wire system. Similarly, the quad pole configuration can be used for retrofitting a three-phase four-wire system. A three-phase transmission line's total power carrying capacity under the unity power factor can be defined as in (6.18). The power carrying capacity of tri-pole (i.e. X=2) and quad-pole (X=3) can be derived from (6.9), as in (6.19) and (6.20), respectively.

$$P_{HVAC} = \sqrt{3}V_{line-to-line}I_{line} \tag{6.18}$$

$$P_{Tri-pole} = 2V_{dc}I_{dc} \tag{6.19}$$

$$P_{Quad-pole} = 3V_{dc}I_{dc} \tag{6.20}$$

Assuming that the retrofitting is performed in such a way that $V_{line-to-line}$ is equal to V_{dc} and I_{line} is equal to I_{dc}, (6.18) becomes (6.21).

$$P_{HVAC} = \sqrt{3}V_{dc}I_{dc} \tag{6.21}$$

Thus, when compared to (6.21), the power capacity of the tri-pole is 1.5 times, and the same of the quad-pole is 1.7 times greater than the HVAC system.

6.5 Proposed Controller

Fig. 6.14 shows the MMC controller with proposed changes. The dotted boxes with t_d and switches S_2, S_3, S_4, S_5 show the proposed changes in the controller.

6.5.1 Conventional controller

In an HVDC network, one MMC controls the power flow over the dc lines while the other maintains the dc voltage. As shown in Fig. 6.14, with switch 'S_1' is at P^*_{Ti}, the MMC operates in power flow control mode, while with 'S_1' switched at P^*_{VTi}, the MMC is in the dc voltage control mode. The P^*_{Ti} is an active power reference defined by the operator, while the P^*_{VTi} is the power reference derived by the voltage control. As depicted, the reference P^*_{VTi} and the arm switching functions (S_{uTi}, S_{lTi}) are derived using the transmission line current and voltage measurement as feedback.

To establish the requirement of the modified controller, a tri-pole configuration is used, as shown in Fig. 6.15, for fault analysis. The current and voltage plots for a dc line to line fault between line 1 and line 2, near terminal 1 are shown in Fig. 6.16. Following the fault, the pole voltage (V_{L1}), and pole-to-pole voltage (V_{dc11}) of MMC11, start falling. The dc current I_{dc11} increases more than 10 times. The SM capacitors u_{c11} get discharged and contribute to the high value of fault current. Similar or even worse observations can be made during other faults. With such distorted feedback, the dc voltage controller and the arm switching function generator yield an inaccurate power reference and switching function, respectively. The controller itself starts feeding inaccurate commands and causes an adverse impact.

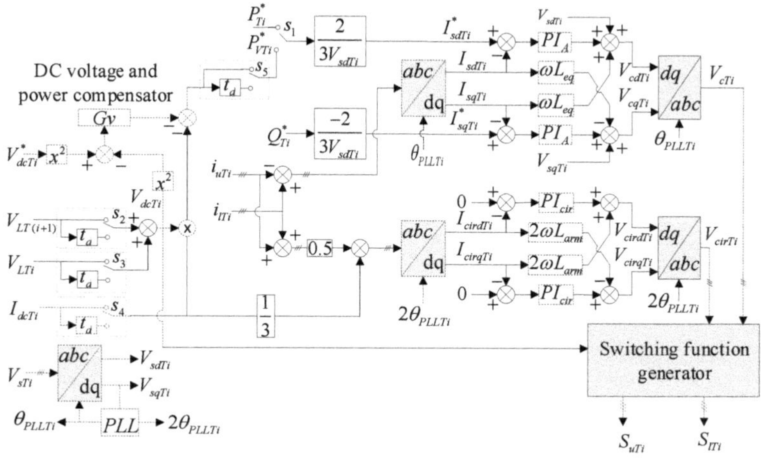

Fig. 6.14. Proposed controller

6.5.2 Fault analysis

System behaviour under the fault condition has been analysed for algorithm development. Though the tri-pole configuration of Fig. 6.15 has been referred for the analysis, the observations are applicable to higher-order multi-pole structures too. The different faults considered in this study are:

1. dc single line to ground faults (f_{SLG1}, f_{SLG2}, f_{SLG3})

2. dc line to line fault between consecutive lines (f_{LL12}, f_{LL23})

3. dc line to line fault between alternate lines (f_{LL13})

In the multi-pole MMC HVDC, when a line to ground fault (f_{SLG}) occurs at a particular line/pole, the voltage of that line/pole reaches zero. The voltages of other line/poles shifts to higher or lower values to maintain the reference pole to pole voltage. Though the circuit is not complete, and the fault current cannot flow, the sudden shifts in the pole voltages cause voltage fluctuations at both ac and dc terminals. It leads to current and power oscillations as well. Line 1 and line 3 are connected to positive and negative poles, respectively. Line 2 is at the middle pole, which can be at zero or very small voltage based on operating conditions. Thus, a fault at line 2 (i.e. f_{SLG2}) to the ground causes a negligible disturbance and can be ignored. However, a fault at line 1 or line 3 (i.e. f_{SLG1} or f_{SLG3}) to ground demands

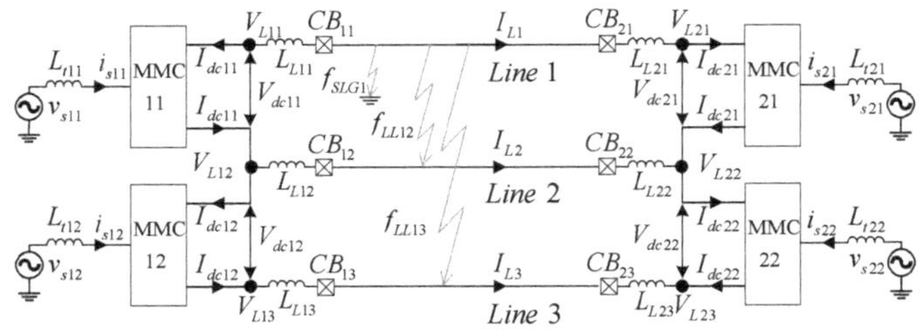

Fig. 6.15. Even symmetrical tri-pole architecture for fault analysis

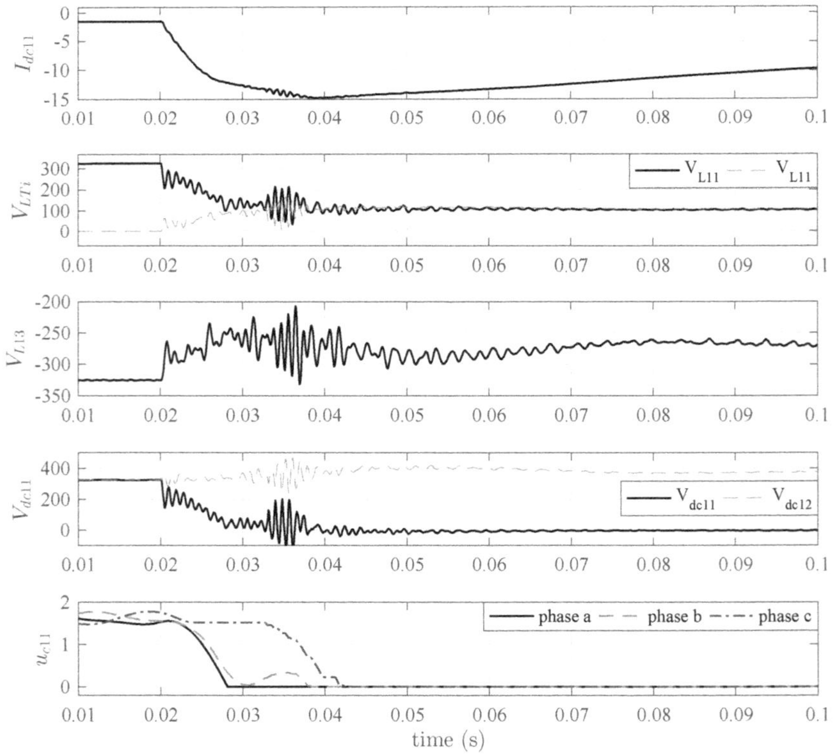

Fig. 6.16. Impact of short-circuit between line 1 and line 2

a dedicated control strategy to minimise the oscillations.

A dc line to line fault (f_{LL}) between consecutive lines is primarily a short circuit between the two poles of the MMC. It leads to the system shutdown as HBSM lacks the dc fault handling capability. Even if the fault gets cleared and the system is restored, it needs a black start. Based on the lines being shorted, the fault can be defined. For the tri-pole configuration, a fault between line 1 and line 2 is termed as f_{LL12}, and similarly f_{LL23}, for fault between lines 2 and 3. During the fault, the pole to pole voltage (V_{dcTi}) between faulty poles goes towards zero, and the current shoots up as the SM capacitors start discharging through the fault circuit.

In the multi-pole configuration, the fault could be between the alternate lines too. In tri-pole, a short-circuit between line 1 and line 3 (f_{LL13}) connected to positive and negative poles of the HVDC terminals, respectively, may occur. It causes voltage collapse, and massive fault current that may damage the converter components or the transmission line itself.

6.5.3 Proposed controller and control algorithm

The proposed controller, as shown in Fig. 6.14, uses four extra switches (S_2, S_3, S_4, and S_5) with a delay block to replace the distorted measurements with the healthy ones. Healthy measurements are the delayed version of the measurement taken a few cycles earlier when the system was under healthy operating conditions. The present study considers a delay of 0.2 seconds for delaying the healthy measurements to replace the real-time measurements under fault. This temporary replacement makes the controller insensitive to the distorted current and voltage feedback under fault conditions and avoids adverse control commands. Further, the circuit breakers are operated in such a manner that the faulty section is isolated while ensuring the power flow over the dc lines is not compromised. Fig. 6.17 presents a flow chart of the proposed control algorithm. Based on fault type, it activates a dedicated command sequence. The possible dc fault can be categorised as follows:

- Any single line to ground fault (f_{SLG})

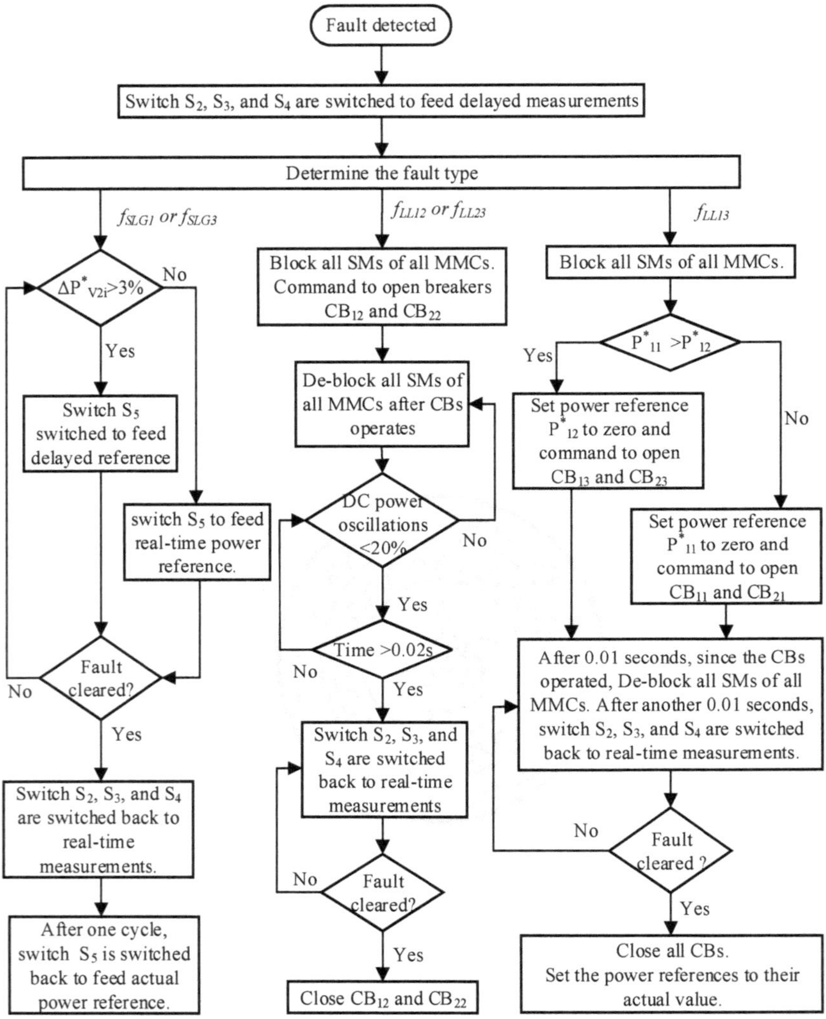

Fig. 6.17. Flow chart: FRT algorithm

- Any line to neutral (middle) line (f_{LL12} and f_{LL23})
- Between high voltage lines (f_{LL13})

As discussed previously, the single line (line 1 or line 3) to ground fault does not contribute any short-circuit current but the dc and ac variables observe fluctuations.

The fault between line-2 to the ground does not require any action. When a fault occurs, switches S_2, S_3, and S_4 are switched to feed the healthy current and voltage measurements were taken a few cycles earlier when the system was in normal condition. This would ensure the proper power references to the MMCs operating under dc voltage control mode and the switching function of all the MMCs. Further, based on the fault type, dedicated steps are followed.

(i) Single line to ground fault (f_{SLG})

As discussed earlier, it has been achieved by making the controller insensitive to the distortion in the feedback signals during the fault. The steps for fault ride-through (FRT) are as follows:

(a) During the fault, the derived power reference (P^*_{V2i}) is compared with the power reference derived a few cycles earlier. Whenever, the difference ΔP^*_{V2i} is more than 3%, the switch S_5 is asked to feed the delayed healthy reference.

(b) As soon as the fault is cleared, the S_2, S_3, and S_4 are switched back to feed the real-time measurements.

(c) Sometimes, the ΔP^*_{V2i} is more than 3%, even after the fault is cleared. In such cases, the S_5 is set to feed the delayed healthy power reference for an additional one cycle. After one cycle of fault clearance, S_5 is switched back to feed real-time power reference.

ii) Line to neutral fault (f_{LL12} and f_{LL23})

The complete algorithm can be summarised as follows:

(a) As soon as the fault is detected, all the SMs of all the MMCs are blocked. As the HBSM does not possess the dc fault blocking capability, non-blocking of SMs would lead to high fault current, as the SM capacitor would discharge through the faulty circuit. A command is sent to open the circuit breakers CB_{12} and CB_{23} for disconnecting line 2.

(b) The dc circuit breakers are expected to operate in 3 milliseconds. As soon as the breakers disconnect the line carrying the difference as well as the fault current, from the MMCs perspective, the fault is cleared and the SM are instantly de-blocked.

(c) Though the faulty section is removed, its impact persists a bit longer. The dc line voltage, current, and so the dc power witness oscillations. The proposed algorithm measures the oscillation peaks of the dominant sub-harmonic components and as soon as it reduces below 20%, the switch S_2, S_3, and S_4 can be switched to feed real-time data. However, a minimum duration of one cycle (i.e. 0.02 seconds) is set to feed the healthy measurements, even if 20% limit is achieved within one cycle.

(d) Once the fault is cleared, the circuit breakers are closed to reinsert the line to carry the difference current.

iii) Between high voltage line (f_{LL13})

The complete algorithm can be summarised as:

(a) When the fault f_{LL13} is detected, all the SMs of all MMCs are blocked. With SM capacitor isolated, it helps in limiting the fault current.

(b) Power references of MMCs in power control mode (i.e. P_{11}^* and P_{12}^*) are compared. The lower one among both is set to zero, and the respective line is disconnected by opening the circuit breakers.

(c) Half a cycle (i.e. 0.01 seconds) after the breaker operation, the SMs of all the MMCs are de-blocked. After another half cycle, the switches are reset to feed the real-time measurements. Now only two transmission lines are transferring the power and the third one is out of the circuit. This reduces the net power transferred by the HVDC.

(d) After the fault is cleared, the circuit breakers are closed to reinsert the line and the power reference is set to its original value.

6.6 Results and Discussion

The performance of the proposed controller and its algorithm is verified on a tri-pole MMC-HVDC configuration (Fig. 6.15) using the real-time digital simulator (RTDS). The ac grids are to be strong enough to sustain the power transients of a few milliseconds (ms) and able to maintain the point of common coupling (PCC) voltage. The MMCs at terminal 1 are controlling active power flow while the MMCs at terminal 2 are maintaining the dc voltage of the HVDC lines. The MMCs are designed for 500 MVA power with 166 kV line to line ac voltage and 320 kV pole to pole dc voltage. Both terminals are 100 km apart, and the HVDC is transmitting the power from terminal 1 to terminal 2. Each transmission line has a resistance of $0.032\Omega/km$.

First, the performance analysis for change in power flow and operation under different voltages is carried out. For the dc fault analysis, (1) single line to ground fault on line 1 (f_{SLG1}), (2) line to line fault between line 1 and line 2 (f_{LL12}), and (3) line to line fault between line 1 and line 3 (f_{LL13}) are considered. The fault location is considered to be near terminal 1.

6.6.1 Active power change

Fig. 6.18 presents the plots of symmetrical tri-pole HVDC during the change in active power. Initially, each MMCs of terminal 1 are drawing 400 MW of active power from the ac network and feeding into the dc lines. The power delivered to ac grids connected to MMCs at terminal 2 is 385 MW each. The reactive power exchange with the ac network is 0 MVAr for each MMC. The current in line 1 (I_{L1}, shown by a dash-dot line) and in line 3 (I_{L3}, shown by a dashed line) are 1.225 kA and -1.225 kA, respectively. Line 2 is carrying the difference current (I_{L2}, shown by

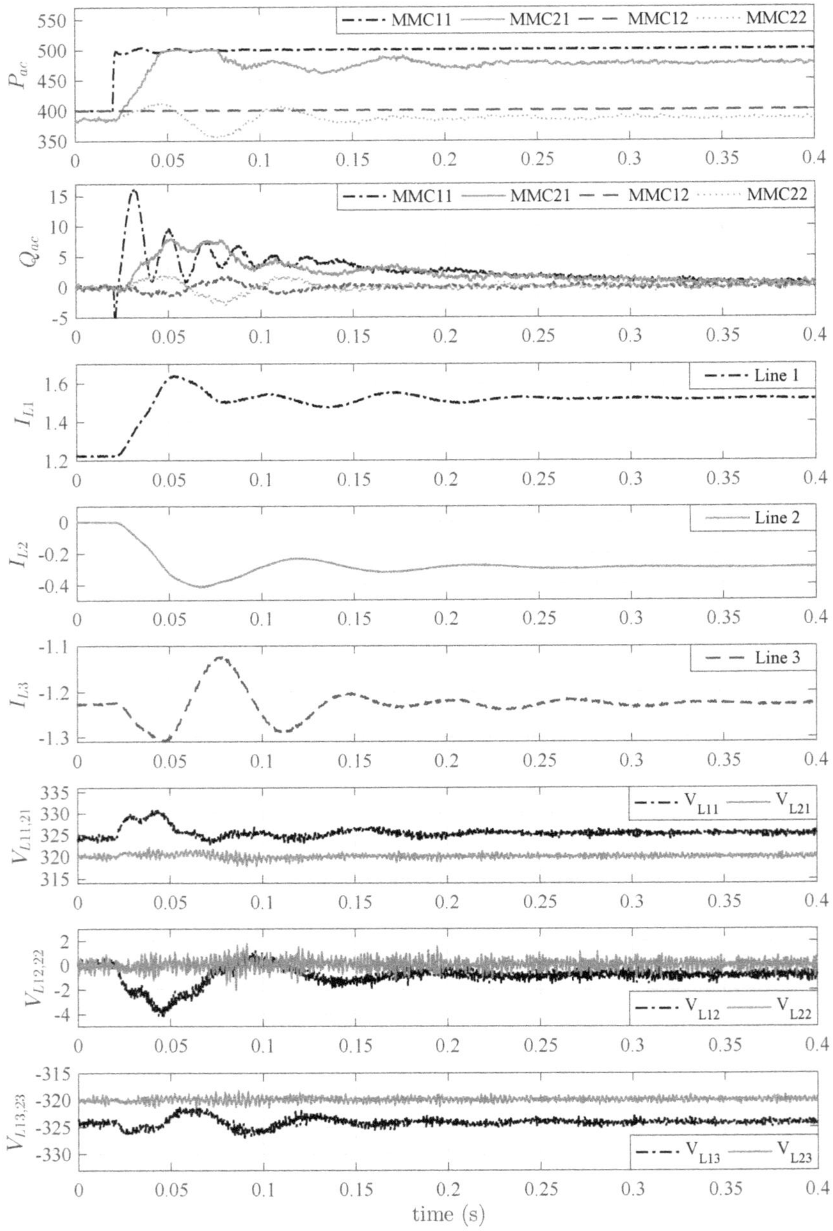

Fig. 6.18. Change in active power flow

solid line), which is 0 kA.

Terminal 2 is in voltage control mode and ensures the rated 320 kV pole to pole voltage. The two voltages V_{L21} and V_{L11} across line 1 are 320 kV and 325 kV, respectively. The difference between the two voltages is the line drop. Similarly, in line 3, the pole voltages V_{L23} and V_{L13} are, respectively, -320 kV and -325 kV. In line 2, the current flow and the respective voltage drop are zero. Thus, both pole voltages V_{L22} and V_{L12} are at the same value, i.e. 0 kV.

At the time instant of 0.02 seconds, the power reference of the MMC11 is increased to 500 MW. Following the change, the MMC21 increases its power delivery. The change in current also increases the voltage drop across line 1 and line 2. The voltage drop in line 2 disturbs the voltage equilibrium. The pole voltages V_{L21}, V_{L22}, and V_{L23} stay at the previous values ensuring 320 kV pole to pole voltage. The voltages V_{L11} and V_{L12} increase following the voltage drop, and it causes a shift in V_{L13} too. The change in V_{L12} and V_{L13} causes a change in I_{L3} as well as the active power exchanged by MMC22. The transients in pole voltages result in transients in the active and reactive power as well as in the dc line current. Once the pole voltages of terminal 1 obtain the new equilibrium and the oscillations are died out, the transients in all other variables die too. The line currents I_{L1} and I_{L2} attain final values of 1.516 kA and -0.291 kA, respectively. The current I_{L3} attains its previous value of -1.225 kA. The transients in reactive power reach up to the maximum value of 15 MVAr and 5 MVAr in the case of MMC11 and MMC21, respectively.

Thus, the change in active power flow between one set of MMCs causes the transients in the remaining MMCs. However, the impact is temporary and the multi-pole HVDC attains a new equilibrium within a few cycles.

6.6.2 Single line to ground fault at line 1 (f_{SLG1})

Fig. 6.19 presents the active and reactive power exchange between MMC and the ac grid at the point of common coupling (PCC) during the f_{SLG1}. The dc line currents and terminal 1 dc pole voltages for the same are shown in Fig. 6.20. Initially, the terminal 1 MMCs are withdrawing 500 MW each, from the respective ac grid, while both MMCs at terminal 2 are feeding 478 MW each. The reactive power support for all four MMCs is 0 MVAr. The voltages are set to the rated values. The pole voltages V_{L11}, V_{L12}, and V_{L13} are respectively, 325 kV, 0 kV, and -325 kV. Line currents I_{L1}, I_{L2}, and I_{L3} are 1.52 kA, 0 kA, and -1.52 kA, respectively. The f_{SLG1} fault occurs at line 1, near terminal 1, at 0.02 seconds and sustains for 0.1 seconds (i.e., 5 cycles).

In this case, no fault current exists as the close path for the same is unavailable. However, for the case without the proposed controller, the power plots of each MMC show high transients. It is due to transients in the dc pole voltages. The line to ground fault causes a voltage shift in V_{L11} from 325 kV to 0 kV. This in turn leads to shift in V_{L12}, and V_{L13} from 0 kV to -325 kV, and from -325 kV to -650 kV, respectively. On terminal 2 also, a similar voltage shift is observed. All the three line currents witness a huge oscillation due to the oscillations in pole voltages. The oscillations in power plots are due to the dependency of the controller on the dc pole voltages and line currents. Although the voltage and current are oscillating due to fault, their impacts on the power plots are minimized with the proposed controller. Without the proposed controller, the P_{ac11} reaches a maximum up to 530 MW, and a minimum up to 410 MW. However, with the proposed controller, it reduces within bands of 517 MW and 490 MW. A similar reduction can be observed in P_{ac12}. Oscillations of more than 200 MW in P_{ac21} and P_{ac22}, reduced to lesser than 15 MW. The reactive power oscillations are also reduced by half for the MMCs at terminal 1 and the same for terminal 2 is one-fifth. The fault is cleared at 0.12 seconds. However, the switch S_5 for MMC21 switched back to feed the real-time power reference derived from dc line parameters at 0.1 seconds (one cycle before the fault is cleared), as the error in power reference ΔP^*_{V21} becomes less than 3%. On MMC22, this happens at 0.12 seconds, at the same time when the fault is cleared and switches S_2, S_3, and S_4 are switched back. The switching causes small transients of negligible magnitude and dies out quickly. When the fault is cleared, all the pole

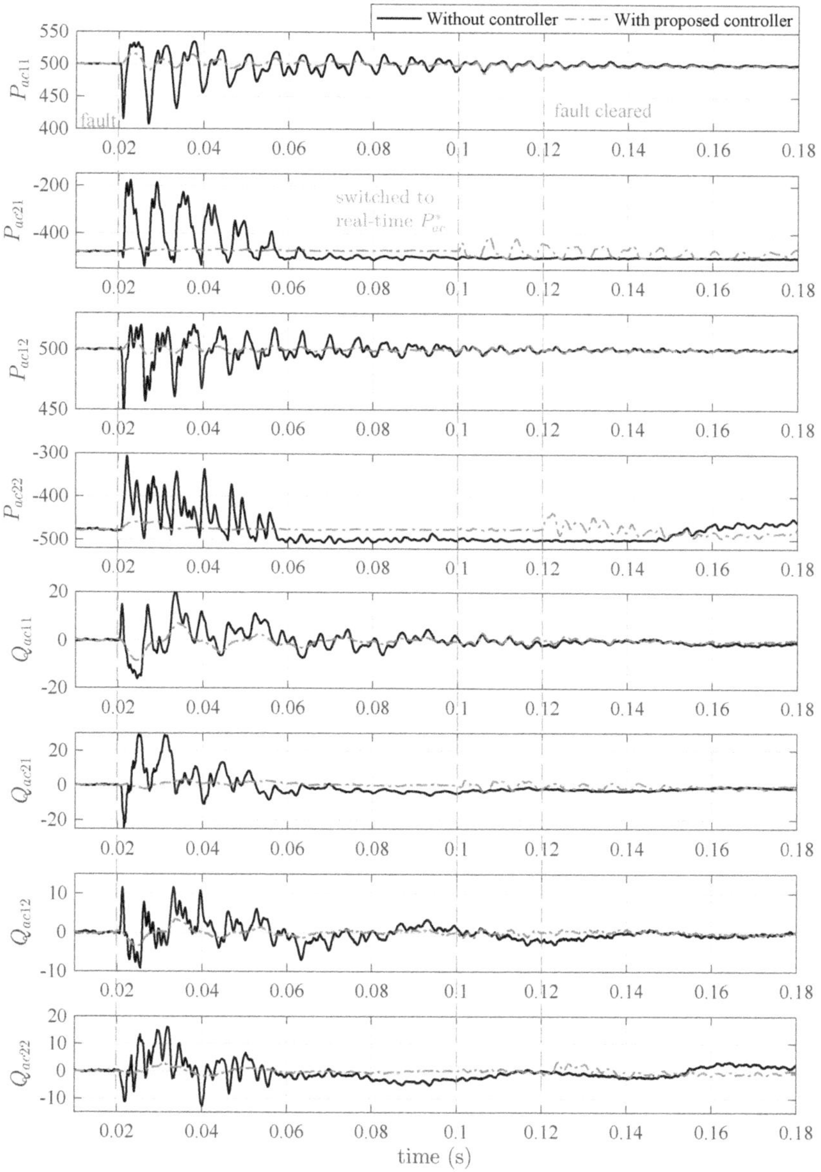

Fig. 6.19. Active and reactive power exchange between MMC and the ac grid during SLG fault

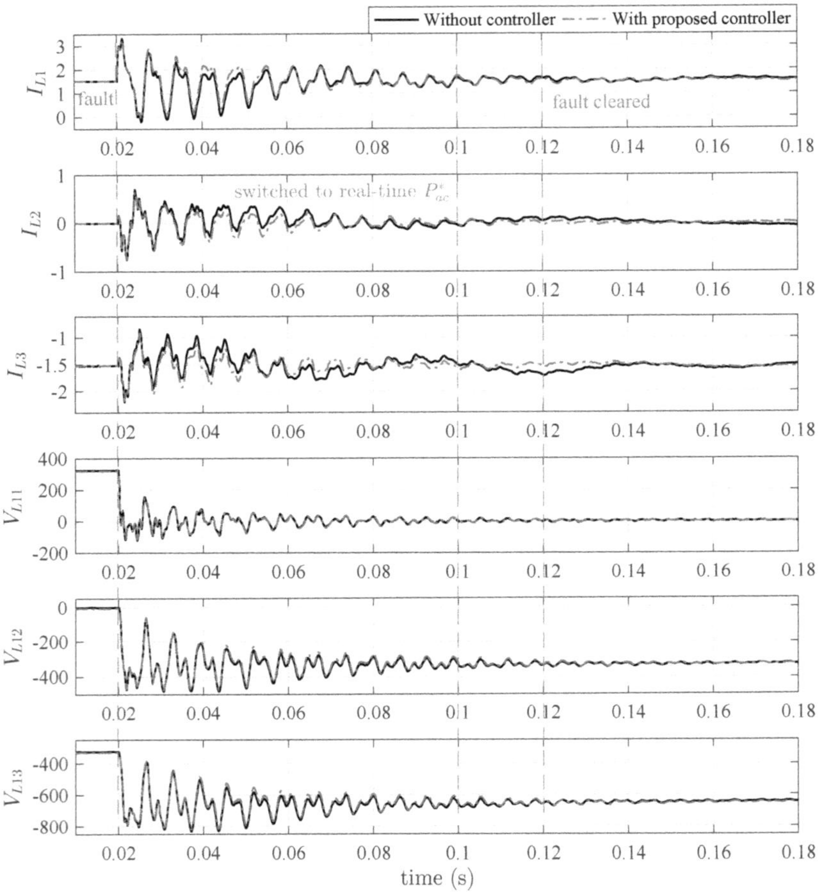

Fig. 6.20. dc line currents and terminal 1 dc pole voltages during SLG fault

voltages gradually move towards their pre-fault values.

Fig. 6.21 and Fig. 6.22 present the SM capacitor voltage (u_{c11}) of MMC11, and upper arm current (i_{u11}) and (i_{u21}), respectively of MMC11 and MMC21 of the tri-pole operating without and with proposed controllers, respectively. The comparison of the two cases shows a shift in voltage due to the reactive power transients when the controller is not applied. However, with the proposed controller, the shift is eliminated and the SM capacitor voltage does not show any disturbance, as the transients in reactive power are suppressed. The arm currents are distorted in both

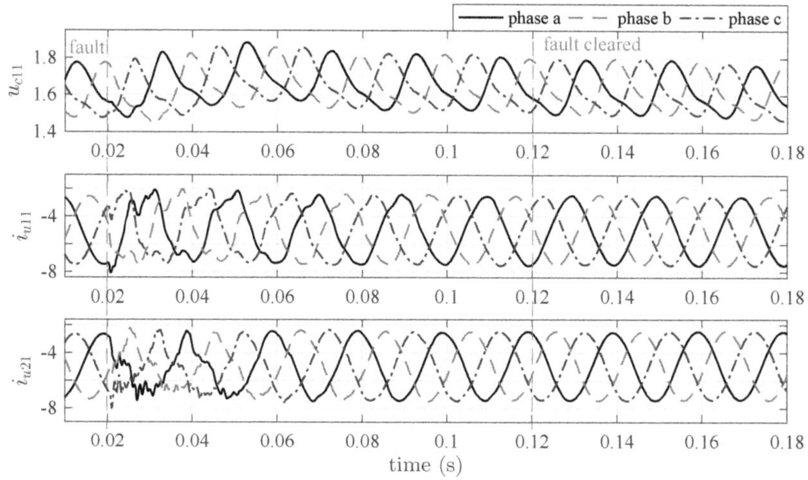

Fig. 6.21. SM capacitor voltage and arm current during SLG fault without proposed controller

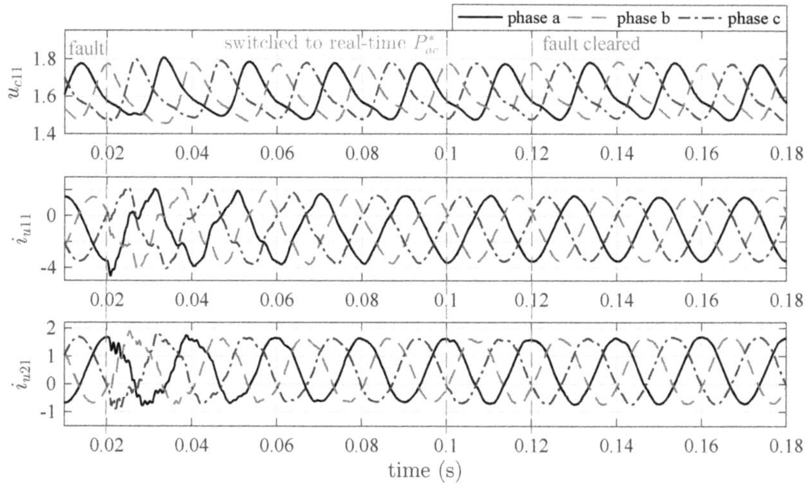

Fig. 6.22. SM capacitor voltage and arm current during SLG fault when the proposed controller is applied

cases. However, the proposed scheme eliminates the distortion at a much faster rate.

6.6.3 Line to line fault between line 1 and line 2 (f_{LL12})

As discussed earlier, when the f_{LL12} fault is detected, a command is sent to open the circuit breakers CB_{12} and CB_{22}. This will disconnect line 2 from the tri-pole. Line 2 is selected specifically, as its isolation would convert the tri-pole into a bi-pole without any compromise in the power transfer capability of the network. The CB operation is not an instant operation and may take 0.003 seconds to operate. This delay causes huge oscillations in the network. The proposed algorithm minimises the oscillations. Fig. 6.23 shows the plots of active and reactive power exchanged by MMCs at the PCC. The line current and pole voltage are presented in Fig. 6.24. The initial operating conditions are identical as in the previous case. The f_{LL12} occurs at 0.02 seconds.

As shown in Fig. 6.23, when the f_{LL12} occurs, the peak values of P_{ac11} and P_{ac13} reach to 1100 MW and 900 MW, respectively. However, with the proposed algorithm, the peak values are reduced to 800 MW and 0 MW, respectively. At the receiving terminal, P_{ac21} and P_{ac22} observed oscillation of huge magnitude across the zero axes. It sustains up to 4-5 cycles and gradually shifts towards the 500 MW. With the proposed algorithm, the oscillations are suppressed within a half-cycle (i.e., 0.01 seconds). At the instant 0.12 seconds, when the switches are switched back, the power flow observes a slight nudge, and the reference value is restored.

Fig. 6.24 presents the plots for line currents and pole voltages for terminal 1. At the instant 0.02 seconds, when the fault f_{LL12} occurs, the current in both lines starts increasing rapidly, while the dc voltage V_{dc11} of MMC11 starts collapsing as pole voltage V_{L11} starts decreasing from its normal value of 320 kV to 0, and the V_{L12} is started increasing from 0 kV (i.e. pre-fault level). The voltage controllers of both MMCs try to maintain the dc voltage. The cumulative effect leads to oscillations in V_{L11} and V_{L12}. With the fault detection, the SMs of all the MMCs are blocked as part of 1st response. This restricts the fault current as the SM capacitors are out of the circuit. The currents in line 1 and line 2 reach only up to the 5 kA and -4.5 kA respectively. However, if the MMCs are not blocked, the fault current reaches

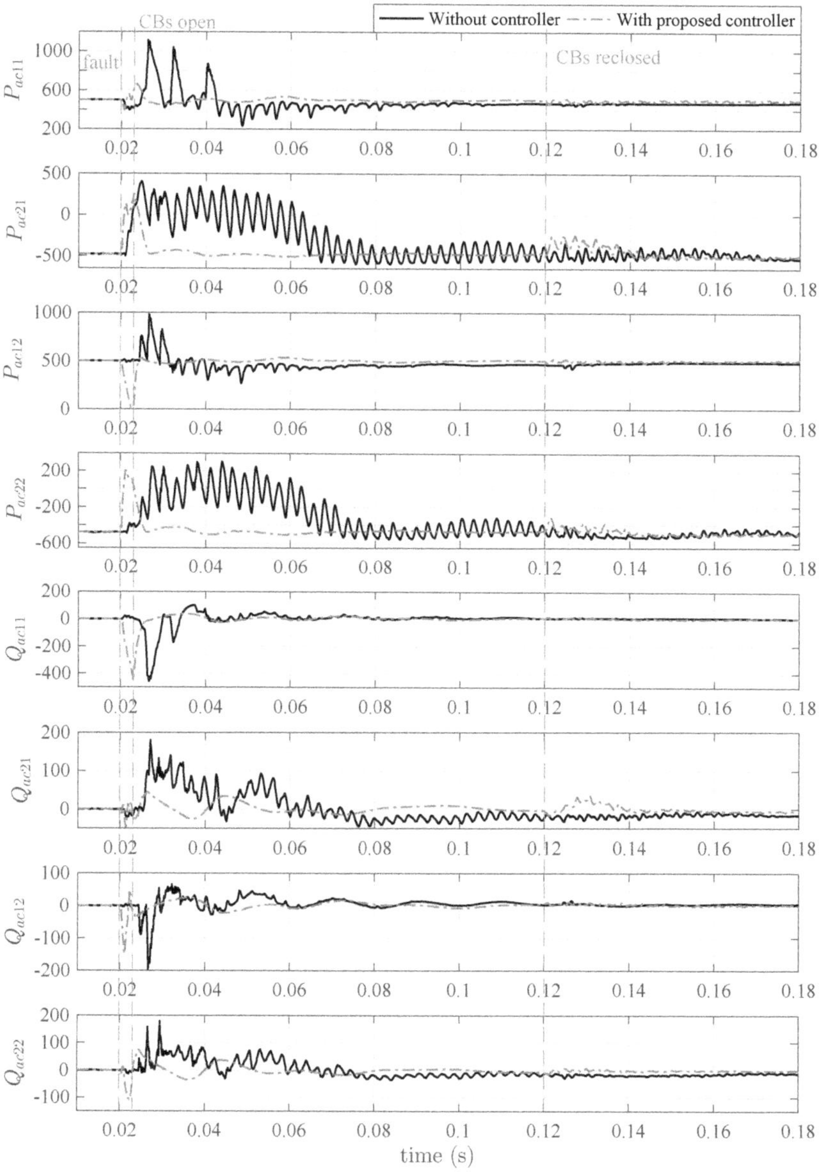

Fig. 6.23. Active and reactive power exchange between MMC and the ac grid during LL12 fault

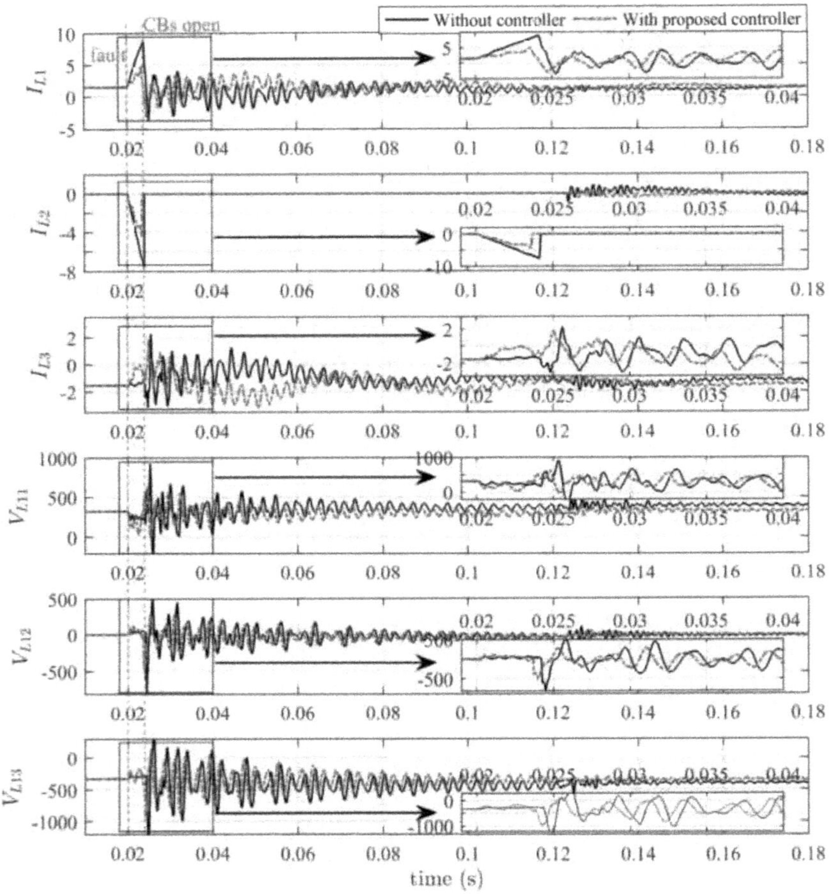

Fig. 6.24. dc line currents and terminal 1 dc pole voltages during LL12 fault

up to -7.5 kA. At the instant 0.023 seconds, the breakers CB_{12} and CB_{22} operate and disconnect the line 2. Since the network does not have any fault now, the SMs are de-blocked. The de-blocking of SMs leads to severe oscillation on pole voltages. However, the proposed algorithm of bypassing the real-time measurements through S_2, S_3, and S_4 reduces the peaks of oscillations significantly.

Fig. 6.25 and Fig. 6.26 present the upper arm SM capacitor voltage of MMC11 and upper arm current of MMC21 when the tri-pole is operated without and with the proposed controllers. It can be seen that with the proposed algorithm,

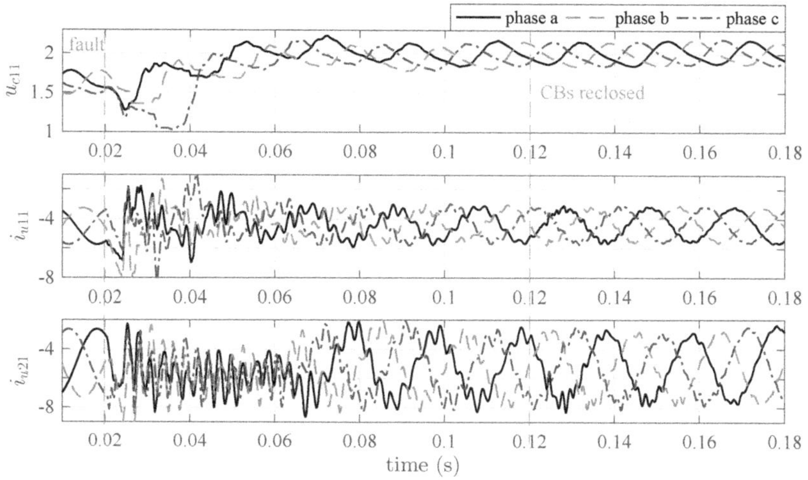

Fig. 6.25. SM capacitor voltage and arm current during LL12 fault without proposed controller

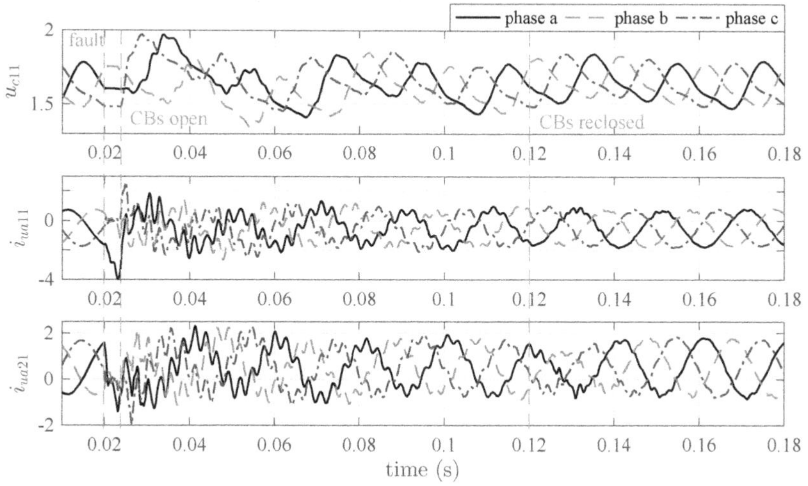

Fig. 6.26. SM capacitor voltage and arm current during LL12 fault when the proposed controller is applied

the reactive power oscillations are reduced significantly. Thus, the tri-pole configuration with the proposed control and algorithm effectively performs the fault-ride through during f_{LL12}.

6.6.4 Line to line fault between line 1 and line 3 (f_{LL13})

The controller performance is also tested for a fault between line 1 and line 3 (f_{LL13}). In the present study, line 3 has been considered for the disconnection of fault. Fig. 6.27 presents the power exchange between the MMCs and the ac network. The dc line current and voltage are shown in Fig. 6.28. The CBs are capable of breaking a current up to 10 kA, within 0.003 seconds.

The initial operating conditions are the same as in the previous cases and a fault f_{LL13} occurs at 0.02 seconds. As shown by the dashed line, following the fault, the transient in power P_{ac11}, reaches beyond 1000 MW. At terminal 2, while the P_{ac21} witness a power reversal, the P_{ac22} initially becomes zero. When the fault is cleared at 0.12 seconds, it also witnesses a power reversal. The fault currents in lines 1 and 3 rise very fast due to short-circuiting between poles of opposite polarity. It reaches beyond 10 kA well before the CB operation. Following the large magnitude of the current, the pole voltages collapse and observe huge oscillations. However, with the proposed algorithm, the CB_{13} and CB_{23}, successfully disconnect line 3. Though the CB disconnects line 3 in 0.003 seconds, the MMCs remain blocked for another 0.01 seconds. At 0.03 seconds, when all the MMCs are de-blocked, the current I_{L1} tries to attain its pre-fault value. At 0.04 seconds, the switches S_2, S_3, and S_4 are switched back to feed the real-time measurements. As shown in Fig. 6.27, the transients in P_{ac11} and P_{ac21} are relatively low with proposed controller. They attain their pre-fault values within a half-cycle. While P_{ac12} and P_{ac22} become zero, as the line 3 is disconnected, power reference is also zero. Similarly, the oscillation in the reactive power is reduced significantly.

At 0.12 seconds, when the fault is cleared, the proposed controller immediately connects line 3. This reflects a small oscillation in pole voltages which are harmless and die out within 2-3 cycles. Without the proposed algorithm, the tri-pole requires complete shutdown and black-start. Fig. 6.29 and Fig. 6.30 present upper SM capacitor voltage and upper arm currents of MMC11, and MMC22, respectively, without and with the proposed controller. With the proposed controller, the SM

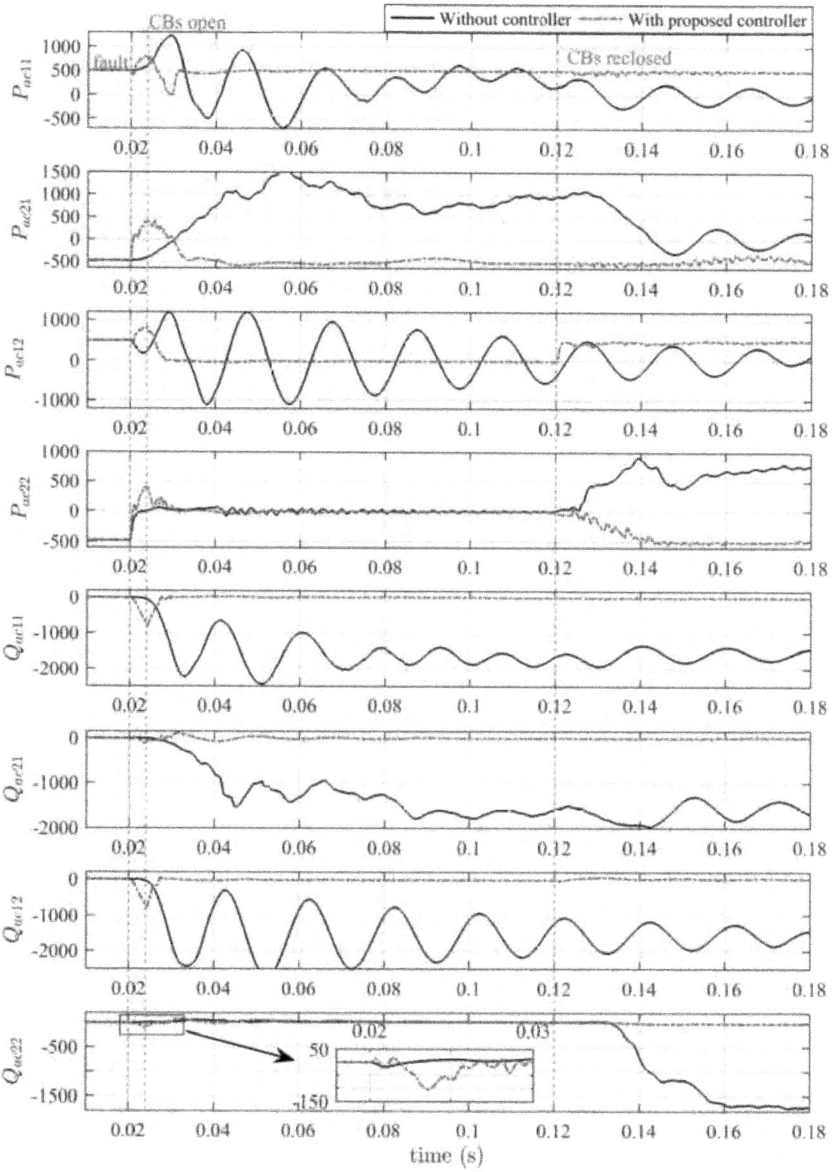

Fig. 6.27. Active and reactive power exchange between MMC and the ac grid during LL13 fault

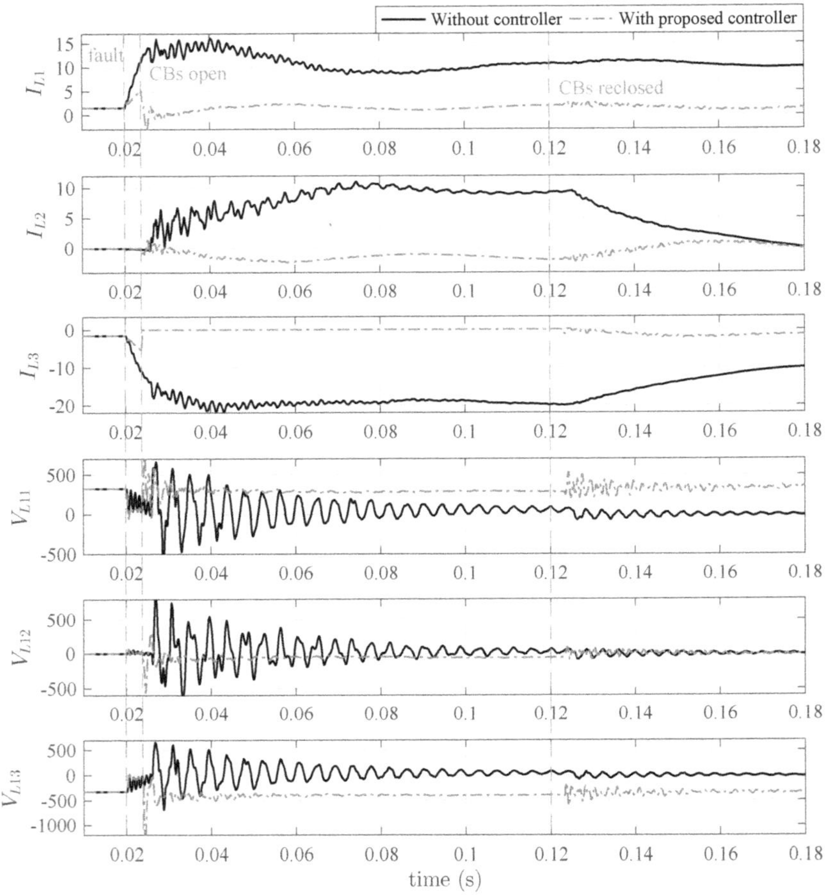

Fig. 6.28. dc line currents and terminal 1 dc pole voltages during LL13 fault

capacitor voltage oscillates within a band and settles over time. Similarly, the arm currents are distorted for half-cycles and regain quickly to their normal value with the proposed controller whereas, it reaches up to eight times their normal value without the proposed controller.

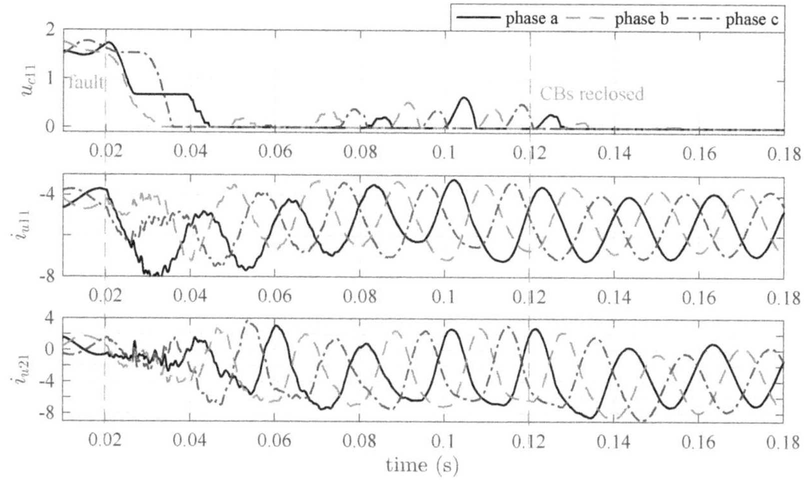

Fig. 6.29. SM capacitor voltage and arm current during LL13 fault without proposed controller

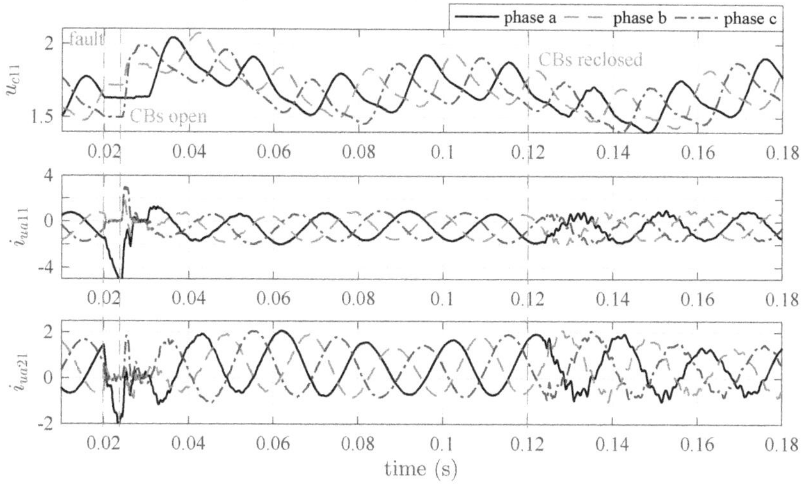

Fig. 6.30. SM capacitor voltage and arm current during LL13 fault when a proposed controller is applied

6.7 Conclusion

This chapter analyses the symmetrical and asymmetrical tri-pole configuration. The MMCs are used at one terminal while the second terminal is simulated using a constant dc voltage source. Further, the symmetrical multi-pole HVDC with MMC converters at both terminals has been studied. It can be sub-divided into even and odd symmetrical multi-pole configurations. These configurations can be used to upgrade a three-phase ac network into a three-pole dc network.

With identical MMCs and operating conditions, individual converters of symmetrical configuration operate in a unipolar mode. The middle pole observes 0 kV, while the voltage stress at the remaining poles is of twice magnitude. The current in the line connected to the middle pole is the difference current, equal to 0 kA. Likewise, converters of asymmetrical configuration operate in a bipolar mode. In this case, the dc poles are at nominal voltage. However, the current in the is line shared by both MMC and attains a value twice the current in other dc lines. The analysis shows that when one terminal is simulated as the constant dc source, the MMC variables are entirely insensitive to change in the active or reactive power on any other MMC for both symmetrical and symmetrical configurations. The change in voltage affects all-pole voltages. However, with the MMCs at both terminals, the entire system becomes sensitive to the slightest change. A controller with the dc fault ride-through capability has been proposed. Tri-pole is a special case of the even symmetrical multi-pole. The same is used for testing the proposed controller and its algorithm.

The conventional controller philosophy has an adverse impact on the system during fault conditions. The proposed controller temporarily bypasses the faulty measurements with healthy measurements taken a few cycles earlier. The fault analysis has been carried out for the single line-to-ground fault and line-to-line fault on the dc side. The performance analysis shows that the transients during line to ground fault are suppressed significantly with the proposed controller. Under the line-to-line fault between the adjacent lines, the proposed logic effectively converts the tri-pole architecture into a bi-pole and reduces oscillations. In case of line-to-line fault between alternate lines, the controller successfully isolates one of the lines and re-instates the power flow through the remaining structure. The proposed controller successfully incorporates dc fault ride-through capabilities into the HBSM-based multipole MMC HVDC transmission.

Chapter 7

Conclusions

7.1 General

MMC HVDC transmission is the latest and most suitable choice for bulk power transfer over long distances. The MMC has evolved remarkably in recent years. Globally, multiple MMC HVDC transmission systems have been installed with 1 GW capacity, while many with 2 GW capacity are under development. MMC is a variant of voltage source converter (VSC). Various advantages of the MMC comprise modularity, scalability, reduced losses, high power quality, reduced operating cost, reduced cost of maintenance, and filter-less and transformerless operations. The sophisticated control that considers concurrent regulation of (i) power flow, (ii) dc voltage control, (iii) eliminating the circulating current, (iv) capacitor voltage balancing, (v) active-reactive power decoupling, (vi) dc-fault handling, etc., is a major challenge in MMC operation. Furthermore, the efficiency is heavily dependent on modulation and submodule (SM) selection techniques. The existing literature of the ongoing research work on MMC includes submodule topology, electromagnetic transient (EMT) model, control strategies, etc. However, there still exist various challenges and opportunities.

Few of them have motivated the work presented in this thesis. The MMC operation is an efficient operation of SM distributed in six arms. The search for optimal SM topology is still going on. Further, the MMC suffers from the flow of circulating current. The existing controllers are not able to facilitate power decoupling. Transformerless MMC HVDC and multipolar MMC HVDC are few among the unexplored application of MMC. They have the potential for minimising the cost and increasing the power transfer capability of the MMC HVDC transmission systems. The present work has focused on designing and analysing the controllers for different aspects of half-bridge SM-based MMC HVDC transmission. The main contributions of the thesis are as follows.

- The architecture, operation and the various SM topologies have been thoroughly reviewed.
- An EMT and small-signal model of MMC has been developed, and an investigation of the significance of the circulating current controller in the operation and stability of the MMC.
- Analysis of internal dynamics and development of the detailed model of MMC have been performed. Three controllers based on internal dynamics have been proposed for the efficient operation of MMC and active-reactive power decoupling.
- The transformerless operation and its impact on different aspects of MMC HVDC systems have been analysed. Two controllers are proposed to eliminate the third-order zero-sequence current, ensuring stable and efficient operation. The dynamics equations have been updated to accommodate the elimination of the transformer.
- The symmetrical and asymmetrical multipolar MMC HVDC transmission have been studied. Further, the symmetrical configuration has been explored to derive even and odd configurations. The impact of change in operating conditions has been analysed. Further, the control algorithm for the dc-fault ride-through capabilities of HBSM-based tri-pole MMC HVDC transmission has been proposed.

The aim of this chapter is to highlight the main findings of the work carried out in the thesis and provide new suggestions for future work.

7.2 Summary of Important Findings

In **Chapter 2**, a comprehensive review of MMC topologies and submodules is offered to provide an understanding of emerging technology. It gives details about the MMC architecture and operation. The new variant of the voltage source converter is a combination of submodules (SM), that are evenly distributed across the six arms. The operation is essentially a voltage synthesis process by controlling the modes of SMs. The process of finding optimal submodule structure has resulted in numerous topologies, and study on the subject is ongoing. These SMs are categorised into various groups based on their voltage levels. The mathematical relationships between capacitor current and SM voltage and the switching signal have been described. In addition, the key issues and challenges of MMC are briefly reviewed.

- The half-bridge submodule (HBSM) is the most commonly used SM topology. This two-level SM is economical and simple in operation but lacks the dc-fault handling capability. It has been used in existing MMC HVDC projects across the globe.

- Full-bridge SM is the next most preferred SM. This three-level SM can generate the reverse voltage and has dc-fault handling capability but is relatively costlier and has higher losses.

- Other SM topologies are primarily modifications or combinations of these two basic SMs. Few special SM for specific applications such as inductive power transfer, current source converter, and hybrid MMC are also present in the literature.

In **Chapter 3**, the MMC's small-signal and EMT models have been developed using RTDS and MATLAB, respectively. The developed EMT model considers 200 half-bridge submodules in each arm of the MMC. The EMT model is validated in comparison to the small-signal model. To analyse the impact of circulating current, two configurations are studied: with and without a circulating current controller. The controllers are developed using the conventional method. The stability analysis, including the eigenvalue and root-locus, has been performed. The investigation of stability limits with variation in controller variables has been carried out over a wide range. A study of the necessity and impact of the circulating current controller (also known as auxiliary controller) on the dynamics of the MMC is presented. The eigenvalue demonstrates that the presence of a circulating current controller improves the system's stability.

- The auxiliary controller reduces the circulating current by 200 folds. It reduces the voltage stress, power losses, and cooling requirements.

- The developed small-signal models are of the 24th and 20th orders, for with and without controller mode, respectively.

- Eigenvalue analysis reveals that most of the modes are substantially stable and damped. Only a few modes (i.e. 3rd, 5th, 6th, and 7th) associated with the SM capacitor voltages are poorly damped. The addition of auxiliary control improves two of the poorly damped modes (3rd and 5th) while having a marginal detrimental effect on two others.

- Eigenvalues of the 2nd, 5th, and 6th modes shift away from the imaginary axis.

In **Chapter 4**, the internal dynamics-based detailed model for the fundamental and circulating current of the MMC has been developed. The models are developed in the d-q domain using the dynamic phasor approach. MMC's modular architecture distributes the capacitors uniformly over its six arms leading to a complex internal dynamic that must not be overlooked. However, the conventional approach for the controller design ignores it. The detailed modelling demonstrates additional cross-coupling between the d- and q-axes, as well as an inter-coupling between the fundamental and circulating current models. Three controller sets are proposed considering the detailed model: detailed, mixed, and simplified. In a detailed controller, both primary and auxiliary controllers take into account all inter- and cross-couplings. Then, by merging a detailed primary and a conventional auxiliary controller, a mixed controller is developed. This control structure considers all coupling except primary to auxiliary inter-coupling. Excluding the remaining inter-coupling results in a simplified version that only considers cross-couplings. MMC models and controllers are simulated on RTDS for steady-state. The comparative performance analysis of proposed controllers demonstrates that the proposed simplified controller outperforms the other control methods.

- Inter-coupling from the primary to the auxiliary controller causes oscillations in the detailed controller.

- Inter-coupling from auxiliary to primary has no substantial advantage in a mixed controller due to its small magnitude. The inter-coupling components are less than 2.5% of the respective signal of the primary controller.

- All three controllers give better decoupling of active and reactive power than the conventional controller.

- The simplified controller is free from the drawback of oscillation. It also reduces the voltage stress on the SM capacitors.

Chapter 5 successfully analyses the transformerless MMC using the real-time digital simulator (RTDS). The research demonstrates that the internal dynamic of MMC causes a third-order zero-sequence harmonic current during this phase. This current passes through the MMC, the ac source, and the dc line and can cause significant damage. It causes an adverse impact on system operation, energy consumption, and power transmission capability. The dynamic equations have been modified to accommodate the transformerless mode. Two distinct controllers are proposed to eliminate the unwanted third-order current: proportional resonant (PR) and proportional-integral (PI). The electromagnetic transient models of MMC under different configurations have been developed on the real-time digital simulator. Moreover, the transformerless MMC with a tertiary PI controller, a 22nd order state-space model, has been developed. The utilities can greatly reduce the cost of the converter station with the proposed configuration.

- In transformerless mode, the third-order zero-sequence currents in MMC arms and dc lines reach up to 15 A and 45 A, respectively.
- Elimination of transformer leads to dc power oscillation up to 14.4 MW.
- The impact on arm energy requirement is linked with the phase of zero-sequence current with respect to fundamental current.
- Both of the proposed controllers efficiently eliminate the third-order current and do not affect any internal or external variable adversely.
- The implementation of tertiary controllers enhances the stability of the system.

Chapter 6 presents the analysis of the symmetrical and asymmetrical tripole configurations. MMCs are employed at one terminal, while a continuous dc voltage source is used to imitate the second. Furthermore, the symmetrical multi-pole HVDC with MMC converters at both ends has been investigated. It has even and odd symmetrical multi-pole layouts. These configurations can be used to transform a three-phase alternating current network into a three-pole direct current network. A controller with the ability to ride through dc faults has been proposed. The tripole, a subset of the even symmetrical multi-pole, is utilised to test the proposed controller and its algorithm. During fault conditions, the conventional controller philosophy has a negative impact on the system. The proposed controller temporarily replaces the faulty measurements with healthy values acquired a few cycles earlier. On the dc side, fault analysis is performed for a single line-to-ground fault and a line-to-line fault. The proposed controller successfully incorporates dc fault ride-through capabilities into the HBSM-based multipole MMC HVDC transmission.

- Under identical operating conditions, individual converters of symmetrical configuration operate in a unipolar mode. The current in the line connecting the middle poles is 0 kA. The voltage stress at the outer poles doubles while the centre poles observe 0 kV.

- The converters of asymmetrical configuration operate in a bipolar mode. When the dc poles are at nominal voltage, the current in the shared line is doubled.

- When one terminal is emulated as a constant dc source, the MMC variables are completely unresponsive to the changes in any other MMC's operating parameters, except for the change in dc voltage. However, with MMCs at both ends, the entire system becomes sensitive to even the smallest change.

- The proposed controller suppresses transients during a line-to-ground fault. It efficiently transforms the tri-pole architecture into a bi-pole architecture to safeguard the system and reduces the oscillations during a fault between adjacent lines. When a fault occurs between alternate lines, the controller successfully isolates one of the lines and restores power flow through the remaining structure.

7.3 Future Scope of Work

The growth of MMC HVDC transmission is in its initial phase. With the growing need for energy, it offers prominent opportunities that can be explored in future work. The extended work of the presented research in this thesis on controllers for MMC HVDC transmission can be as follows:

- The work on the literature review can be extended to include a comparative analysis of switching and conduction losses of the SMs. Further, the hybrid MMC alternatives also should be explored.

- In future research work, circulating current controller may be considered for MMC operation under an unbalanced ac grid or unbalanced dc poles

- Similarly, the research work on the internal dynamics-based controller can be done to extend the scope of the controller for the unbalanced grid scenarios. The different faults on the ac terminals of MMC can be considered.

- The transformerless operation of MMC HVDC can be extended for the multi-terminal MMC HVDC transmission network. Further, the impact of faults on the ac grid also can be analysed.

- The techno-economic analysis of transformerless MMC for different HVDC applications could be an area of investigation.

- It could be interesting to expand the proposed controller for tripole configuration to the other multipole MMC HVDC combinations. Furthermore, the relationship between delay time and stability should be investigated. The tripole configuration offers an opportunity to upgrade the existing ac line into the dc lines. A techno-economic and other related studies can be performed to analyse its feasibility.

- The asymmetrical multipole configuration also offers a scope of the investigation.